The Lighter Side of Rail:
A Guide to Open Deck Bridges

Namita

Contents

1 Preamble

1.1 Book layout

Chapter 1: Literature Review

The literature review provides an understanding of railway sleepers and highlights the difference between open an closed deck bridges. Special focus is placed on investigating the nature of railway loading and resulting effects on precast concrete sleepers. A global shift away from the limit stress design and toward a limit state philosophy has highlighted the need to gain a better understand of prestresssed concrete sleepers response to both the static and especially the fatigue loading. Methods of fatigue testing in the lab and modelling fatigue using numerical assistance is also presented.

Chapter 2: Methodology

Guided by the literature review, the Australian standard (AS 1084.14-2012) was used to provide a basis for the physical test procedures. S-N curves, from the fib 2010 model, were used as a basis for fatigue life predictions.

Chapter 3: Finite Element Tuning

The commercial finite element software, Abaqus, was used to build a numerical model of the low profile sleeper. The constitutive equations were studied to find which parameters significantly influenced the finite element outcomes. Once these parameters had been identified they were tuned using findings from the static test.

Chapter 4: Results and Discussion

The load vs displacement and load vs strain which were obtained from the physical tests are displayed. Using the numerical model stresses in both the steel and concrete were obtained in addition to the displacement and strain. These stress vs load relationships allow the applied load to be related to a fatigue life using the aforementioned S-N curves.

Chapter 5: Conclusion

The findings of the study are presented.

1.2   Introduction

Railway sleepers or railway ties as called in America are integral components of a railway track. They provide the multi-purpose function of maintaining the rails level and at an equal gauge distance along the track while transferring the wheel load to the sub-base or supporting structure.(Manalo et al., 2010) When the track passes over level terrain the sub-base usually takes the form of a layer of ballast supported by layers of graded soil. Elevated track as found on railway bridges can be divided into two categories,depending on the deck structure; open deck or closed deck. An open deck track as illustrated by Figure 1.1 is supported by transverses sleepers known as transoms. Transoms are connected to longitudinal steel girders which form part of the bridge structure. Open deck bridges are almost exclusively used on steel bridge structures.

Figure 1.1: Open decked bridge

Due to the lack of ballast, opened decked bridges support less dead load and can thus be more slender than closed deck bridges. Additionally there is no drainage problem or large areas for snow to build up as opposed to closed decks. For these reasons open deck bridge are cheaper options and have become the most common bridge system (Griffin et al., 2014). Historically timber has been exclusively used as the transom material, however the wooden transoms are vulnerable to fire and biological attack thus they need frequent maintenance. Due to the lack of ballast, track alignment is problematic over open decked bridges. There is also a significant change in track stiffness from a bridge track to the ballasted approach track. The change in track stiffness reduces the ride smoothness and results in accelerated damage to multiple components of the track and train.

Closed deck bridges as illustrated in Figure 1.2 have a solid deck on which either standard ballast track is layered or the track is directly fixed to continuous slabs without the use of a ballast layer. Closed deck bridges are generally supported by concrete or masonry structures due the heavier deck structure. Closed deck bridges provide superior riding quality and are especially suited to high speed and heavy haul track as the ballast provides better vibration damping and relatively smooth stiffness transition from the approach tacks. Direct fixation of the track to the deck provides the ability to reduce the rail height, an important consideration where minimizing vertical clearance is desired such as in tunnels. Although initially more expensive than ballast, direct fixation has a reduced maintenance cost and track down time as the track alignment is not affected by ballast wear and settlement.

Figure 1.2: Closed deck bridge (Moreu et al., 2014)

1.3   Problem description

South Africa has an extensive yet ageing rail network. The majority of South African Railway bridges are open decked bridges with wooden sleepers. Wooden sleepers having a design life of 10 years require regular replacing. South Africa does not have adequate resource to supply the replaced sleepers locally and thus relies on imported timber sleepers. Due to the world wide shortage of hardwood, importing these sleepers is both costly and unsustainable.

Concrete sleepers have successfully replaced wooden sleepers on ballasted track but to date have not be used on open decked bridges. The behaviour of low profile sleepers when subjected to a range of loading patterns is not well understood or documented. Before a statement regarding the suitability of low profile concrete sleepers replacing wooden sleepers on open decked bridges can be made a better understanding of the sleepers needs to be gained.

1.4   Aims

This project aims to develop an understanding of low profile concrete sleeper's response to both static and fatigue loading. Specific aims are:

1. The mode of failure should be identified for both static and fatigue loading

2. The ultimate static load capacity should be found

3. The fatigue life of the sleeper should be presented

1.5   Scope and Limitations

While there are potentially numerous methods of solving the 'wooden sleeper' problem replacement with precast concert sleepers will be investigated exclusively. After studying the literature it became evident that alternative sleeper materials such as glass reinforced polymers seem to offer an ideal alternative to wood. Although the use of these materials has been used to solve a similar problem in Australia and Japan, the lack of long term durability and load response studies leaves the suitability of polymer sleepers as questionable. Polymer sleepers will not be investigated in this study.

South Africa is one of the leading countries in concrete sleeper technology and the vast majority of the country's sleepers are concrete thus it is deemed more valuable to advance our concrete sleeper technology rather than implementing a new technology with little local investment.

The manner in which the sleeper is supported and loaded in the laboratory is not fully representative of that found in the field. A railway has many components such as, rails, rail pads and connection mechanisms. The variation in these components results in large variation in the nature and magnitude of loading that the sleeper experiences.

Figure 1.3 illustrates some the numerous rail components, each component acts as parameter in determining the global railway load response.

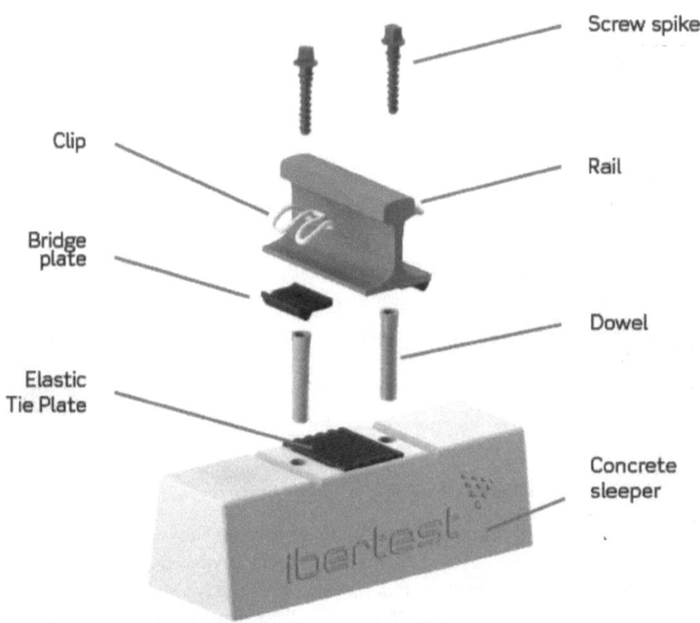

Figure 1.3: Railway components associated with precast concrete sleepers (Ibertest, 2018)

The Australian code (AS 1084.14-2012) will be used to guide the laboratory test set up. In this test the sleeper is tested without the installation of any other railway components. Open decked sleeper are supported on wide girders rather than the pin supports found in the laboratory. Due to the higher bearing pressures and extremely rigid support the laboratory set-up is considered to be a more conservative set up than would be found in the field.

While these deviations from field conditions will not provide field representative findings the study results in the tests being repeatable as they are conducted in accordance with a recognised standard.

The fatigue life of any specimen is dependent on the stress range, the stress schedule and the material deterioration due to environmental factors. In-terms of the fatigue life study, only the effect that stress range has on the sleeper will be investigated. The stress schedule or environmental deterioration factors will not be taken into account. Although stress reversal fatigue could occur on a bridge as a results of track uplift it will not be investigated.

2 Literature Review

2.1 Types of Sleepers

2.1.1 Timber

While South Africa has largely replace wooden sleepers with concrete ones, historically hard wood has been the sleeper material of choice for the vast majority of railways. Timber is an ideal sleeper material and hardwood sleepers are still ubiquitously used in the Unites States of America(Ferdous and Manalo, 2014). Wood has many favourable characteristics such as its easy workability, it is lightweight and is not brittle. Wooden sleepers however have poor durability characteristics and are highly susceptible to microbial attack, fire and wood splitting, consequently wooden sleepers have a relatively short life span of 10 years (Manalo et al., 2010). This results in many millions of timber sleepers needing to be replaced every year. The number are staggering, the US replaces 140 thousand timber sleepers every year and "India imports 7 million timber sleepers per year to maintain her rail network"(Manalo et al., 2010). In an effort to reduce microbial attack, timber sleepers are treated with a toxic chemical know as creasote. The end of life processing for such treated sleepers poses an environmental problem. The declining quality of timber sleepers is also an issue. As timber reserves are depleted, less favourable species and sizes are being used. Due to the light weight of wooden sleepers they have a low lateral stability compared to concrete sleepers consequently train speeds are generality limited to a maximum of 100 km/h (Manalo et al., 2010). Figure 2.1 illustrates the typical dimensions of wooden sleepers in relation to other railway components.

Figure 2.1: Wooden sleepers on open deck bridge bridge (Kalbaskraal Bridge Western Cape South Africa

2.1.2   Steel

Steel sleepers have limited use world wide for a number of reason. Steel sleepers are light and thus proved reduced stability of the track. For this reason train speeds over steel sleepers are generally limited to 160 km/h(Manalo et al., 2010). Although steel sleepers have a quoted design life in excess of 50 years, this is extremely dependent on the aggressiveness of the environment. Their use is limited to dry climates such as inland Australia where they account for 13% of counties sleepers(Manalo et al., 2010). Figure 2.2 illustrates a typical steel sleeper. Depending on the fluctuating steel price, steel sleepers are generally more expensive than other alternatives. Steel sleepers have a major advantage of being able to be 100% recycled where as all other sleeper materials are down cycled or disposed.

Figure 2.2: Steel sleepers (Ste, 2018)

2.1.3 Precast Concrete

Concrete sleepers have enjoyed widespread use since the 1950s and have been used to replace deteriorated wooden sleepers as well on new track segments. This is predominantly due to their long design life of 50 years(Kaewunruen and Remennikov, 2009a). Precast concrete sleepers typically use concrete strengths between 40MPa-60 MPa and are pre tensioned with top and bottom reinforcement wires. Although both steel and concrete are energy intensive materials to produce,Griffin et al. (2015) states that precast concrete sleepers require six times less green house gasses than wooden equivalent over a similar design life period.

Concrete sleepers weighing between 285 kg and 200 kg per sleeper are significantly heavier than wooden sleepers. Consequently concrete sleeper placement requires lifting machinery where as wooden sleepers can be manually placed. The increased weight however offer superior lateral and longitudinal track stability(Manalo et al., 2010)(Miura et al., 1998). Due to this increased stability concrete sleepers are ideally suited to high speed and heavy haul lines. Due to better durability and larger load bearing capacity precast concrete sleepers have largely replaced traditional hard wood sleepers, as an exception, wooden sleepers are still exclusively used on open decked bridges. Tradition concrete sleepers are deeper that their wooden counterparts. Retro-fitment of concrete sleepers, using a deeper section causes interference with existing services and existing railway levels. This is a predominant factor which inhibits traditional concrete sleepers directly replacing existing wooden sleepers. Low Profile concrete sleepers are a speciality sleeper which have been designed to have the same vertical dimensions as traditional wooden sleeper and can thus be used to replace wooden sleepers where height restriction is an issue. Figure 2.3 illustrates the dimensional differences between traditional concrete sleepers, Low Profile sleepers and wooden sleepers.

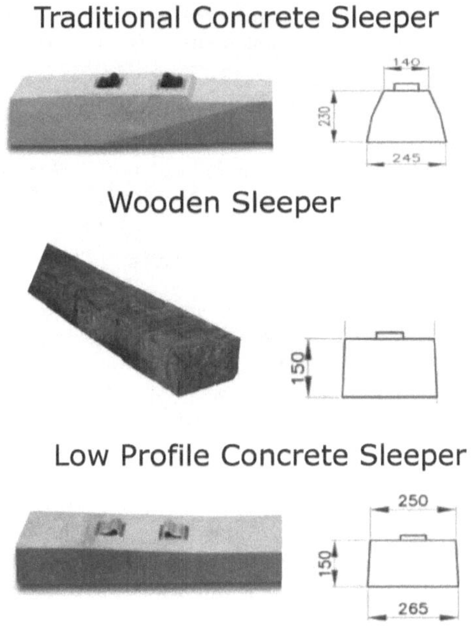

Figure 2.3: Comparison of different sleeper dimensions

According to Ferdous and Manalo (2014) South Africa has approximately 43 million sleepers in service and requires 305 thousand new sleepers per year, of which the vast majority are concrete. Although monoblock sleepers are the most prolific type of sleeper, other types of concrete sleepers are used in specific areas of the rail network; Twin-bloc and ladder sleepers being examples. Ladder sleepers can reach 12 m lengths and are designed as follows; two longitudinal concrete beams continuously support the rail and are connected at intervals by steel ladder rungs. Ladder sleepers reduce ballast deterioration and provide smooth transfer to track with a change in sub base stiffens such as bridge approaches (Manalo et al., 2010). The Universal Concrete sleeper developed by Infraset is commonly used to replace wooden sleepers in rail turnout sections.

Pre-cast concrete sleepers have been used successfully on open deck railways as illustrated by (Abo et al., 2008) where heavy snowfall in Japan necessitated the use of open decked bridge. However the long-term functioning of this bridge has not been assessed thus this instillation cannot be used as a case study.

Concrete sleepers are extremely stiff compared to wood and they thus do not have the same damping characteristics as wood. Concrete being an inherently brittle material results in precast sleepers being susceptible to impact damage, during transport, placement and use. They are also susceptible to rail seat abrasion which has resulted in some catastrophic derailments(Erp and Rogers, 2008).

2.1.4 Polymer/Fiber Composite Alternatives

Significant research has been conducted on the possibility of using modern engineering materials such as thermoset plastics, fibre reinforced polymers and recycled thermoplastics in rail sleeper production. Australia and Japan seem to be leading this research. The research is focused on designing a synthetic sleeper with similar mechanic, and workability properties of timber. Manalo et al. (2010) states that the Japanese have developed such a sleeper using polyethylene foam reinforced with glass fibre. This sleeper has a stated 60 year design life. Other materials such as recycled plastic sleepers exhibit favourable mechanical properties and are currently undergoing in-service testing in the USA and India (Manalo et al., 2010). Notably these sleepers have been selected to be used on open decked bridges as an ideal replacement for wooden sleepers. This is due to the fact that they exhibit good damping and impact resistance characteristics.

Fiber-reinforced foamed urethane (FFU) is a synthetic material which has successfully been used to produce railway sleepers. This material is gaining wide spread popularity amounts railway companies, to date 1 300 km of track worldwide are supported by FFU sleepers. The dynamic properties such as impact resistance are similar to wood and the mechanical properties exceed those of wood(Koller, 2015). Koller (2015) conducted a full range of tests on FFU sleepers at the Munich University of Technology which resulted in the German Federal Railway Authority approving FFU for use on its railways.

CarbonLoc TM is a commercial polymer sleeper which is currently being used in Australia as a direct replacement for wooden bridge sleepers. Erp and Rogers (2008) states that CarbonLoc sleepers require 8 times less energy to produce than prestressed concrete sleepers for a similar strength transom. The use of CarbonLoc on a main heavy haul coal line signifies its full adoption within the Australian rail industry.

While Polymer/Fiber composite sleepers have proved to be ideal replacements for wooden transoms their widespread use is limited by their higher cost and a reluctance of rail companies to adopt a product with limited in-service track records.

2.1.5 Slab track

The advent of high speed (faster than 200km/h) trains has resulted in a new track development, that being slab track. A slab track is classified as ballast-less. There are no sleepers as the rails are directly fixed to a solid reinforced concrete slab at regular intervals. The development of slab tracks began in 1965 (Miura et al., 1998). Miura et al. (1998)states that although slab tracks generally cost 1.3 times more than ballasted track, the lower profile and lighter weigh of slab tracks can reduce overall rail cost. For this reason slab tracks are predominantly found along viaducts and in tunnels. Ballast-less track is especially suited to high speed rails as ballast becomes unstable and requires regular maintenance when supporting high speed tracks.

Vibration and noise control is becoming an ever pressing design criteria. Slab tracks have better vibration control potential over ballasted track.(Miura et al., 1998) Although slab track is predominantly use in new track segments, the Sydney harbour rail bridge case study provides a good example of the retro-fitment of slab track. The ageing wooden sleepers used on Sydney harbour bridge required urgent replacement.Griffin et al. (2014) concluded that a modulate track slab was the most feasible replacement option. A slab was installed as it was shallower (180mm) than standard concrete sleepers and thus did not cause interference with existing services.

2.2 Railway Loading

The movement of trains imparts a complex array of loads on the track and supporting structure. Xe et al. (1994) states that railway loading can be divided into two distinct loading categories. The first being quasi-dynamic loading caused by a undamaged rail wheel rolling along a track. Quasi-dynamic is defined as a force that is time dependent yet slow enough to ignore the inertia forces of the system components. In numerous studies this type of loading is modeled as a sinusoidal force exerted on the supporting sleepers.

The second loading type which is superimposed on top of the quasi-dynamic load is impact loading. Impact loading can be due to wheel flats which are local deformations on the wheel cased by train breaking or track defects such as poor welds or engine burn skids. Impact loading has a short duration. Wheel flats typically cause impact loads with durations of between 1 to 3 micro seconds while track deformities such as bad welds, result in impacts with slightly longer durations between 5 and 10 micro seconds(Xe et al., 1994)(Wang, 1996). Impact loading can cause significant damage due to the high loading magnitudes which can be up to 4 higher than quasi-dynamic or static loading (Kaewunruen and Remennikov, 2009b). Igwemezie et al. (1989) states that the dynamic impact load can be between 200-600 kN while the corresponding static load for a 40 ton axle is only 110kN. As impact loading is defect related it needs to be modelled using statistical occurrence of such loading events. Impact forces have been cited by many researches as a primary reason for pre-cast concrete sleeper damage.

2.3 Impact Loading tests

Drop test are a common method of simulating impact loads caused by train and rail deformities. Drop tests are conducted by raising a specified weight a certain height and allowing it to free fall onto the test subject. The magnitude of the load and duration of the impact is controlled by the drop weigh and the height. A rubber sheet may be placed in between the subject and the drop weight to increase the loading duration(Kaewunruen and Remennikov, 2009a).

Using the drop test Xe et al. (1994) studied the effect that loading rate (impulse force) has on sleeper crack propagation. This study focused on extremely high loading rates, as would be created by a wheel impact load. The impulse force was varied by changing the drop weight and varying the stiffness of the rail pads. A lower rail pad stiffness results in a longer impact time and thus a lower impulse force. Xe et al. (1994) used drop weights of 345kg or 504kg at a height of 1.52m these resulted in impact loads ranging from 343 kN-543 kN depending on the weight and effective stiffness of the system. Many of the sleepers cracked upon first blow. Xe et al. (1994) concluded that as the loading rate increased, the sleepers failure moved from flexural cracking to sheer cracking.

Wang (1996) notes that where the failure mechanism of precast concrete sleepers is flexural under quasi static loading it changes to to shear failure mode under impact loading. This change in failure mode for different loading rates is an important aspect to be noted when testing precast sleepers.

As expected Wang (1996) also found that reducing the concrete strength from 65 MPa to 40 MPa as well as reducing the prestressing level reduced the dynamic stiffness and thus reduced the peak stress associated with impact loading. Lower stiffness rubber pads were found to provide better vibration damping and reduced impact load effects.

In a case study of Canadian railway tracks Igwemezie (1989) found that hairline cracks developed under the loading seat after a few months of operation. The main concern was that the bending resistance would be reduced as the crack widths increased. In a similar but independent study Wang (1996) noticed that precast sleepers were found to be cracked after a few months of in service use. A persistent problem noted by Igwemezie et al. (1989) and Xe et al. (1994) in independent studies was the occurrence of hairline cracks on concrete sleepers on open decked railways. It was suggested that these be the result of impact loading and vibration of the transoms. The cracks developed directly under the rail seat location. These observations led the Canadian Railways to further investigate the effect of impact loads on precast concrete sleepers.

In the late 90's the Canadian railways replaced 22 wooden sleepers with pre cast concrete sleepers along a 8.8 m test bridge. The bridge was a typical open decked steel girder bridge. The sleepers were monitored as live train traffic passed over the rails at different speeds (3.2-96km) and with different wheel conditions.Igwemezie et al. (1989) provides some interesting conclusions from the in-service test, namely: The loading response of sleepers is independent of the train speed when there are no defects on the wheels. Interestingly, for faster moving trains(faster than 70km/h) the maximum strain was found to occur directly under the rail seat rather than at midspan sofit. It was reported that "impact loading transmitted to the ties was independent of the tie support conditions [and that] the rail-tie pad has the greatest influence on the response of the ties to impact loading"(Igwemezie et al., 1989).Igwemezie et al. (1989) concluded that for pre cast concrete sleepers to be used on open decked bridges at least 50 % of the impact load needs to be attenuated, it was proposed that this be done using an improved rail tie pad.

2.3.1 Influence of track parameters

Igwemezie et al. (1989) investigated the loading of the first sleeper on a bridge. This is an important study as the first and last sleeper often experience high loads due to a sudden stiffness change associated with the approach vs bridge support structure. It was noticed that the end sleepers experienced 3-5 times higher loads than the sleepers in the middle of the bridge. This finding corresponds to the fact that sleepers of different stiffness characteristics cannot be used alongside each other, as the sleeper with the higher stiffness will attract more load an may fail prematurely. For this reason piece wise replacement of individual sleepers on a track should be avoided. (Manalo et al., 2010) This investigation suggests what when testing concrete sleepers for the critical loading environment the support condition should be chosen to simulate a sleeper near the bridge abutments. This can be achieved by using a simply supported system with extremely high stiffness supports.

Igwemezie (1989)conducted a detailed parameter study in which various track parameters such as,the stiffness and shape of the sleeper and rail pad,rail weight and sleeper spacing were modified and the corresponding impact loading effect was recorded. Proceeding from this study (Igwemezie, 1989) noted the following observations; "the rail-sleeper pad has the greatest influence on the sleepers response to impact loading". A reduction of rail-sleeper pad stiffness significantly reduces the modes of vibration from the third mode upward. The maximum tensile stress on the sleeper occurs directly underneath the rail seat on the bottom sofit. Rebound cracking proved to cause negative flexural crack at the top of the sleepers. The first five natural frequencies fell between 84 and 950 Hz, this correlated with laboratory research. These observations are considered accurate as similar findings were mirrored by an extensive in field study conducted by (Peters, 1992)

Campbell and Mirza (1984) conducted a similar parametric study to Igwemezie (1989) however the parameters were varied while the wheel load distribution among bridge sleepers was measured. The parameter which had the maximum influences were, the sleeper spacing,the stiffness of the tie girder pad and the the weight of the rail. For a constant rail and pad stiffness the sleeper load was linearly related to the sleeper spacing. These findings were proved numerically and then validated using a physical model. The physical model consisted of only 7 sleepers as the loading pattern for this arrangement of sleepers adequately represented a longer bridge.

While the laboratory and infield studies mentioned above are predominantly related to impact loading they provide valuable insight into the many rail parameters that influence the load resistance of the sleepers. It is expected that that the many rail parameters will have an influence on the fatigue resistance of the sleepers. As this study will not investigate the influence of these parameters it is important to ensure the variable parameters are kept as control variables and that the results obtained are very specific to the set of parameters applicable during testing.

2.4   Response of Bridge Structure

The fundamental equations of motion dictates that the dynamic response of any structure is determined by its stiffness and mass, altering either of these factors will alter the dynamic response of a structure. Replacing wooden sleepers for concrete sleepers will change both the mass and stiffness of the track structure. It is important to understand the effect that this change will have on the dynamic response and interaction of both the supporting bridge and the track itself. Unacceptable vibration of either the track or bridge structure may result if the resonant frequency of the new system happens to move too close to the loading frequency of the passing train or ambient loading such as wind. Low frequency vibrations have a larger displacement amplitude and can cause high member stresses. High frequency vibrations are generally responsible for noise generation and expedited wear and tear of rail components.

The dynamic analysis of a rail bridge is complex as there are many degrees of freedom coupled together. The studies presented below have tried to draw conclusions about the dynamic interaction between the bridge and the track structure.

Igwemezie et al. (1989) observed the dynamic response of a steel bride girder after concrete sleepers had replaced wooden sleepers. The study concluded that girder sleeper pads isolate the bridge girders from vibrations resulting from wheel deformities. (Igwemezie et al., 1989) noted that the the track properties were solely responsible for the track's high frequency response and that under impact loading the sleeper's dynamic response is independent of it's support conditions.

Igwemezie (1989) findings correlate with Cheng et al. (2001) in that they independently found that the "effect of the track structure on the dynamic response of bridge structure was found to be insignificant. However, the effect of the bridge structure on the dynamic response of the track structure is considerable"Cheng et al. (2001).

Campbell and Mirza (1984) concluded that the wheel load distribution among the neighbouring ties was dependent on "the stiffness of the tie-girder pad, type of rail, and the spacing of the ties with little influence on the stiffness of the bridge support. The stiffness and type of support girder was found to have minimal impact on the load distribution between sleepers. This finding reinforces the idea that certain components of the track structure can be studied independently from the supporting bridge structure.

These findings are important to note because they indicate that the dynamic response of the track structure namely sleepers and rails has minimal effect on the dynamic response of the bridge structure. This is not to say that mass and stiffness of the track structure has no effect on the bridge's dynamic response. For as Kim (2010) notes the dynamic vertical displacement of bridge structures is reduced as the mass of the track structure increases. The dynamic response of the track structure caused by train loading will have little effect on the bridge however the dynamic response of the bridge structure will be altered by a change in track type.

2.5 Design Criteria for Sleepers

In many countries concrete railway sleepers are currently designed according to a permissible stress criteria (You et al., 2017). This is a 19th century design method which has predominantly been replaced by limit state methods for most other structural components. The design procedure involves estimating the maximum load the sleeper will support and designing it such that a certain maximum stress is not exceeded in the concrete. For example Transnet specifies that the maximum tensile stress in the sleeper must not exceed 3.5 MPa (Meyer, 2016).The design load is based on a combination of the static load and a dynamic component. The dynamic effect is converted into a quasi-static load by incorporating a dynamic amplification factor. This dynamic amplification factor is normally 3-4 times the static load (Murray and Bian, 2012).

When discussing design criteria it is important to define failure. A sleeper has failed when it can no longer carry out it's purpose, which is primarily to maintain gauge distance and top rail levels. Numerous failure modes exist and are broadly classified into structural failure and non structure failure. The common modes of sleeper failure are presented below:

1. Rail seat cracking-loss in gauge distance (structure)

2. Midspan cracking resulting in large flexural deformations-loss in gauge distance (structural)

3. Midspan cracking allowing extension or gauge spread-loss in gauge distance (structural)

4. Under sleeper abrasion-loss of rail level (non structural)

5. Rail seat abrasion-loss of rail level (non structural)

It is important to note that even when the tensile stress of the concrete has been exceeded and the concrete is thus cracked this does not necessary mean that the sleeper is unable to perform it's function adequately. However according to the permissible stress criteria, cracked sleepers should be replaced even though there may be a significant amount of residual load capacity left. Kaewunruen et al. (2012) concluded that the "permissible stress design criteria underutilizes the sleepers load bearing capacity" and results in uneconomical use of materials. As a result Kaewunruen et al. (2012) concludes that a limit state design method should be adopted. This notion is supported by Murray and Bian (2012) who states "sleepers are generally replaced because of non-design factors such as serious damage due to train derailment".

2.5.1 Limit State Criterion

Limit states are common design criteria in modern engineering design processes. Murray and Bian (2012) presents one of the first limit state design methodologies for concrete sleepers. In this study (Murray and Bian, 2012) uses the Australian structural design code (AS3600-2009) which is a limit state design and aplies the governing principles to determine the limit state of concrete railways sleepers. As identified by Murray and Bian (2012) there are three distinct limit state subcategories; strength,serviceability and fatigue. A comprehensive limit state design should be based on all three sub limit state criteria. The different limit state subcategories are discussed below:

Strength limit state

This limit state is dependent on a once off event that may result in sleeper failure. A probabilistic design approach is required to identify the maximum load the sleeper will experience in a certain return period and thus must be designed to withstand.

The ultimate design load should be specified based on measured in-service loads and a specified design return period. Murray and Bian (2012) recorded a years worth of loading history (2 million reading) as a basis to identity ultimate design load. The ultimate design load is a combination of the normal weight transited through the rolling wheel and the impact loading of deformed wheel. Each loading component has a different statistical distribution of magnitude vs occurrence. The gravitational component was found to have a normal distribution as shown if figure 2.4 while the impact component was found to have a Weibull and Gumbel distribution as shown if figure 2.5. A histogram of the two components is show bellow.

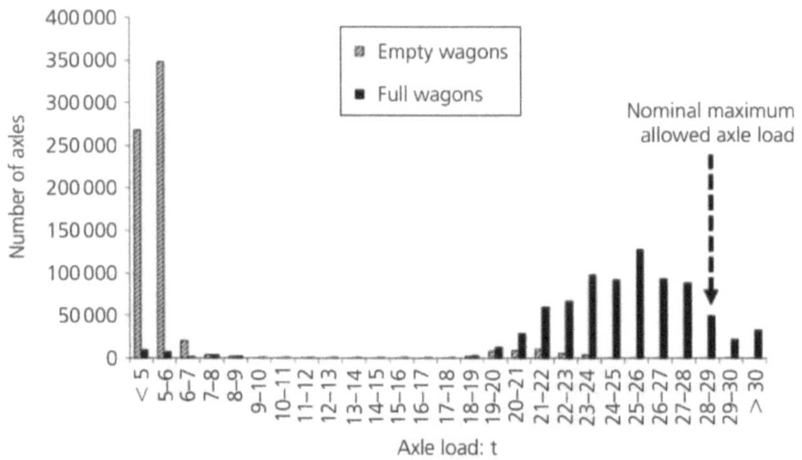

Figure 2.4: Histogram of gravitational loading contribution (unfactored) (Murray and Bian, 2012)

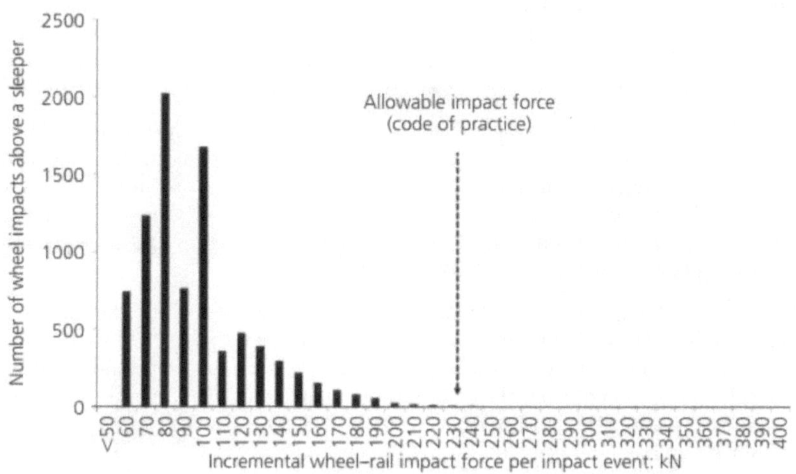

Figure 2.5: Histogram of impact loading contribution (unfactored)(Murray and Bian, 2012)

A 2000 year return period was justified due to the importance of rail networks. Consequently the probability of the design load being exceed once during the 50 year life span of the sleeper is 2.5%. Murray and Bian (2012) concludes with a limit state equation that provides the ultimate design moment as follows.

$$0.8M_u \geq 1.1M_Q + M_I$$

Where M_u stands for the ultimate design bending strength, M_Q stands for the moment induced by the weight of the train factor and M_I stands for the moment induced by the impact load factor. Using the above equation and the historic loading data Murray and Bian (2012) calculated an ultimate design bending moment of $M_u = 49kNm$.

Damageability (fatigue) limit state

The fatigue limit state is a common limit state used in the design of mechanical equipment. Fatigue failure occurs after repetitive loading cycles at loads less than the stress limits would predict. In order to implement a fatigue limit state, the cumulative frequency distribution needs to be know. Figures 2.4and 2.5 are examples of cumulative frequency distributions. Using a suitable fatigue life model such as Miner's hypothesis the fatigue life for the element is calculated for a given design or the element is designed to withstand a certain number of loads in it's design life.

Using the fatigue limit to design sleepers is complex in the sense that there exist multiple modes of fatigue failure. Thus obtaining an accurate fatigue life model that describes these modes accurately is challenging. Obtaining an accurate cumulative frequency distribution for the wheel loads is also challenging as the load experienced by the sleeper is highly influenced by many track parameters and thus recorded measurements are sight specific.(Wakui and Okuda, 1997a)

The scope of Murray and Bian (2012) investigation did not include the fatigue life limit state however he identifies it as an area of further research. A few years later You et al. (2017) conducted a study of the current and past sleeper design design approach and proposed using the fatigue design guidelines found in standard design manuals such as fib 2010 and EN 1992-2:2005 to design sleepers according to a fatigue life limit state.

This approach however is contrary to Wakui and Okuda (1997) who argued that standard fatigue calculations such as those conducted for bridges can not be used as they do no account for impact loads experinced by railways. Wakui and Okuda (1997b) and You et al. (2017) assumed sleepers would fail in fatigue as a result of the steel tendon's failure. No mention was made in any literature of the sleepers failing as a result of concrete fatigue.

Serviceability limit state

Serviceability limits found in existing structural design codes generally include, maximum allowable crack widths and maximum allowable deflections. Although there has not been any explicit research into defining the limit states for concrete sleepers some of the current limits may be applied. For example the precast sleepers are class 1 precast members and thus no cracks are allowed.

Wakui and Okuda (1997a) conducted laboratory tests to quantify the damageability of concrete sleepers. An impact test was conducted 17 thousand times at loads ranging between 210 kN to 330 kN. He concluded that although fine cracks occurred they were within allowable serviceability limits. These observations provide an example of one limit state mechanism (fatigue) being limited by an other limit state namely the serviceability state (crack widths).

This **book** will assist in determining the failure/serviceability life of precast concrete sleepers when used on open decked bridges and loaded under certain conditions, thus similar findings to Wakui and Okuda (1997a) are expected.

2.6 Numerical Models

Numerical models such as finite element models are useful tools to use in modelling track component and their response to train loading. A numerical model's main advantage is that many parameters can be tested without the cost associated with destructive physical tests. ABAQUS, a commercial finite element package will be used in this study.

Earlier fatigue tests conducted on sleepers such as those conducted by Wakui and Okuda (1997b) and You et al. (2017) used standard linear beam theory to predict the stresses in the concrete and steel tendons. The linear equations are unable to capture plastic deformation and thus it is important to employ a finite element model.

As with all models it is vital to choose suitable constitutive material relationships and analysis theories. Numerous studies have confirmed that the Timoshenko beam theory is the most appropriate to model pre-cast concrete sleepers as well as the supported rail Kaewunruen and Remennikov (2009b)(Kaewunruen and Remennikov, 2009a). It is common practice to model the elasticity track components such as rail and girder pads as a series of spring and damper couples. Literature suggests that non-linear solvers should be utilised in the analysis as the rapid loading caused by impact creates non-linear stress distributions within the sleeper. (Griffin et al., 2014)

There are three damage constitutive material relationships available for use in ABAQUS, " (1)Smeared crack concrete model,(2) Brittle crack concrete model and (3) Concrete damaged plasticity (CDP) model"(Wahalathantri et al., 2011). For the modelling of precast concrete in fatigue loading the CDP model is often used as it is able to model reinforced concrete deterioration for cyclic loading.

A brief explanation of each relationship will be provided as a basis for why the CDP model was chosen:

Smeared Crack Model

This model is ideally suited for monotonic loading schedules and is thus not suited for fatigue analysis. Cracks are incorporated via altering the concrete constitutive law for certain elements along the crack path.(Simulia, 2013)

Concrete Damage Plasticity

Utilizes a similar crack propagation procedure as the Smeared crack model however as it "allows the user control of stiffness recovery effects during cyclic load reversals"(Simulia, 2013) this model can be used to simulate concrete deterioration under load reversal.

Brittle Crack Concrete Model

The cracking can only occur due to tensile strain rather than though compressive cracking which CDP account for. The compressive stress strain relationship is assumed to be prefect elastic through the cracking process. (Simulia, 2013)

As the CDP model is not a discrete cracking method, the results will be mesh sensitive. To avoid this it is important to use regularization limiter such as the Hillerborg regularization, if you don't your results will depend on mesh size. (Tao and Chen, 2014)

A numerical model consists of multiple components each with specific parametric values. Examples of model parameters are, effective stiffness of the support restraints, railpad stiffness and sleeper stiffness. It is often impossible to know the exact numerical value of these parameters and thus model calibration is required. A model is calibrated by comparing the physical response to the model's response. The comparative response may be static deflection, dynamic mode shapes or harmonic frequencies. The model parameters are tweaked until the model's response matches that produced by the physical component.

2.7 Fatigue

Fatigue failure was first identified in metallic materials and is subsequently well understood for such materials. Fatigue will be introduced in terms of the general metallic behaviour and will then progresses to the lesser understood fatigue behaviour of concrete. In a classical fatigue model (applicable to metals) the fatigue life of a specimen can be divided into two stages, the crack initiation stage and a crack propagation stage. Most materials have inherit micro defects both internally and on the surface. These defects result in high stress concentrations even when the average stress is well below the elastic limit. Initially the local stress concentrations cause micro cracks. As the loading cycles progress these crack form macro crack, the macro cracks propagate until failure occurs (Ahsan, 2007). Steel and similar materials have a stress limit below which infinite load cycles my be applied without inducing fatigue failure.

2.7.1 Fatigue in Plain concrete

During the hydration and hardening process concrete inherently becomes a cracked material before any loading is applied. The presence of these micro crack reduces the crack initiation stage and thus concrete fatigue is predominantly a crack propagation procedure. Concrete can be considered a three phase material composed from aggregate, cement past and a transition zone (ITZ) which exists between the aggregate and the past. Each component has different mechanical characteristics which result in a complex fatigue failure mechanism. Crack initiation and propagation proceeds at different rates for each of the the different phases. Crack propagation though the ITZ is often cited as the initial location in which crack initiation occurs. (Ahsan, 2007)

A study conducted by Van Ornum (1903) indicates that the fatigue resistance of plain concrete is 50-55% of the static ultimate compression(Fc) after 7 thousand cycles. (Naik et al., 1993) found that the fatigue life of concrete is influenced by intrinsic and extrinsic factors. Intrinsic concrete parameters being ,moisture content,curing method, age of loading, water to cement ratio. As an example of variation the flexural fatigue resistance was found to be between 33% to 64% of FC depending on the moisture content of the sample. Extrinsic factors being loading rate, rest period, loading frequencies and the nature of loading.

Temperature fatigue can also be a cause of fatigue failure however due to the relatively short sleeper length small thermal strains will develop compared to those cause by train loading. Thermal fatigue can thus be ignored for this study. Corrosion fatigue occurs when concrete is subject to cyclic loading in an aggressively corrosive environment. The dual effect significantly reduces the fatigue life of structures.

2.7.2   Fatigue in Pre stressed Concrete Members

Reinforced concrete, being a composite material has an significantly more complex fatigue process than plain concrete. There are three primary failure mechanisms as listed below:

1. Flexural compressive fatigue failure of the concrete. This is a ductile failure as the progressive cracking reduces the stiffness of the beam and results in significant deflections before failure.

2. Fatigue bond slip between the concrete and the reinforcing.

3. Flexural fatigue failure of reinforcing steel. This mode of failure occurs in under reinforce beams. The failure is brittle as the crack progresses through the steel in a rapid and catastrophic nature Ahsan (2007). Over reinforced beams and the presence of compressive reinforcement exhibit a more complex failure mechanism as interaction with the compressive steel and concrete creates complex stress distributions.

4. Shear fatigue failure. This type of failure has significantly lower capacity than the other modes of failure.Olsson and Pettersson (2010) states that shear failure can be as low as 40-60% the static resistance. There are multiple modes of sheer fatigue failure, those being; failure of the sheer reinforcement, longitudinal failure as it crosses the shear crack,compressive failure of the concrete either in the web of the beam or at the top of the sheer crack.(Ahsan, 2007)

Balaguru (1981) conducted a fatigue study on prestressed concrete beams and attributes the increase of crack widths and and reduction of flexural stiffness during fatigue loading to a process of cyclic creep. Cyclic creep takes place in the compressive zone of the concrete and in the the tensile steel causing strain softening of the pre stressing steel. The rate and magnitude of cyclic creep is significantly higher than static creep.

2.8   Fatigue Testing

There are a variety of fatigue tests that have been devised to measure the fatigue life of materials. Two distinct classes of fatigue tests exist; high cycle fatigue and low cycle fatigue testes. A high cycle fatigue test(HCF) refers to a test at low stress levels with a high number of loading repetitions typically $N > 10^5$ while Low cycle fatigue (LCF) occurs when relatively few loading cycles are conducted at high stress level.

It is important to determine which fatigue test best represents the loading environment of the structure. Different tests can produce different failure mechanisms as well as recoded fatigue strengths(Ahsan, 2007). LCF is better suited to seismic loading simulation, whereas HCF better simulates rail loading.

Igwemezie et al. (1989) conducted fatigue tests on two railway sleepers, one had been cracked due to in-service impact loading the other sleeper was taken from the middle of the bridge and was un-cracked. "The cracked sleeper was subject to 21.7 million loads cycles at an equivalent axle load of 350 kN, this is 5 percent higher than the static cracking load" (Igwemezie et al., 1989). No further damage was noticed after 21.7 million cycles. After the uncracked sleeper was subject to a load of 277 kN for 1.87 million cycles a crack appeared however after 20.3 million cycles the crack had not grown and no further damage was visible. Only after increasing the load to 145% of the static load did the sleeper fail at 23.6 million cycles. From these results Igwemezie et al. (1989) concluded that although impacts by wheel defects causes sleepers to crack, the fatigue life is adequate to guarantee sufficient service life. However aggressive environment may cause corrosion and reduce the service life of a cracked sleeper.

High cycle fatigue tests are time-consuming as it can take many days for a million load cycles to be applied to the specimen. Consequently increasing the loading frequency is advantageous as it reduces the testing time. Lloyd et al. (1968) concluded that within the range of 440 cycles/min (7.33Hz) and 70 cycles/min (1.16 Hz) the loading rate did not affect the fatigue resistance of the concrete. These results are compatible with the maximal allowable loading rate of 600 cycles/min (10Hz) as specified in AS 1084.14-2012.

Koller (2015) conducted a range of structural tests on fibre reinforced foamed Urathane (FFU) sleepers, of interest to this book are the fatigue test results.The compression fatigue strength was tested according to DIN EN 13230-2. A 3Hz loading frequency was used with the maximum fatigue load set at 150kN while the lower load was set at 50kN. The FFU sleepers showed no damage after 5 million cycles. A slight deflection increase of 1 mm from (3mm static deflection) was noticed after the first 2 million cycles and a further 0.8mm increase after 5 million cycles. These results proved that the FFU sleeper has excellent fatigue life and are a good reference to which concrete sleeper fatigue life can be compared.

2.9 Fatigue life prediction

Fatigue life prediction can be based on empirical formulation or theoretical formulation. Both the S-N curve and the Goodman and Smith models are common empirical fatigue life prediction modes. Empirical models are limited in their ability to predict failure under complex loading environments. Theoretical formulations have a better ability to model complex loading however they often require the input of many parameters which can be unknown. As this book is based on fatigue strength of concrete sleepers it is important to understand which fatigue life prediction would be best suited to describe the situation. A brief description of each method will be presented below.

2.9.1 S-N curve or Wohler curve

The S-N curve is a material specific empirical model and is commonly used to illustrate the fatigue life of high-cycle fatigue tests. The curve illustrates the number of loads (causing a stress fluctuation) that leads to sample failure. The natural logarithm of the the number of cycles is plotted against the maximum stress of each load cycle. Numerous tests need to be conducted at each stress level in order to obtain statistically acceptable findings.(Ameen and Szymanski, 2006)

The Woher curves presented in the fib Model Code 2010 will be used to predict concrete fatigue. These curves are the results of studies conducted by (Wefer, 2010) on normal, high and ultra high strength concrete.

The fib Model Code 2010 accounts for fatigue deterioration by decreasing the characteristic compressive strength of concrete (Urban et al., 2014).

The fib Model Code 2010 utilities three equations, each applicable for a certain domain, to relate the compressive stress to the number of cycles to failure. These equations and their domains are presented below;

$logN_1 = (12 + 16S_{c,min} + 8S_{c,min}^2)(1 - S_{c,max})$

$$logN_2 = 0.2logN_1(logN_1 - 1)$$

$$logN_3 = logN_2(0.3 - 0.375S_{c,min})/\Delta S_c$$

The following domains apply:

1. if $log_1 <= 6$ then $logN = log_1$
2. if $log_1 > 6$ and $\Delta S_c >= 0.3 - 0.375S_{c,min}$ then $logN = log_2$
3. if $log_1 > 6$ and $\Delta S_c < 0.3 - 0.375S_{c,min}$ then $logN = log_3$

Figure 2.6 illustrates the S-N curve for concrete in pure compression. The 'maximum stress level' is a ratio from zero to one and defined as: the maximum compressive stress experienced in the concrete (MPa) divided by the fatigue reference compressive strength.

The fatigue reference compressive strength is dependent on the age of the concrete before fatigue loading is initiated. Using the equations presented in the fib model it was calculated as 76 MPa for this study.

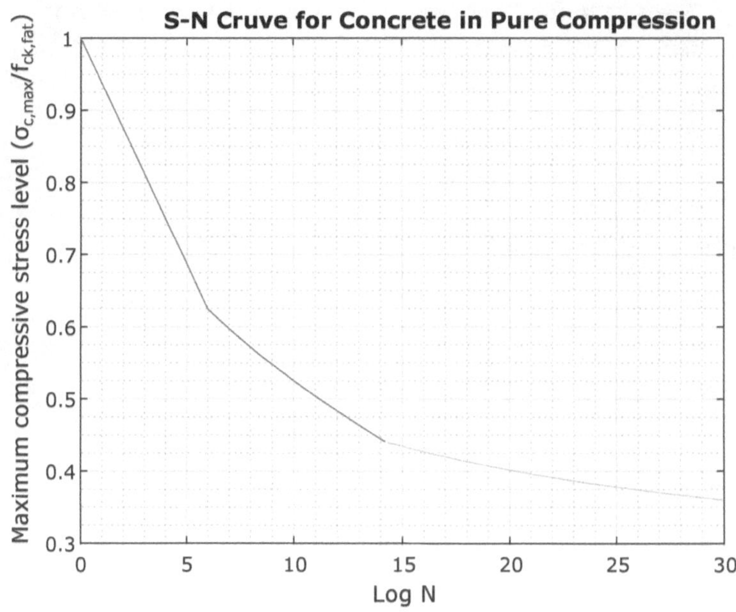

Figure 2.6: S-N curve for Concrete in Pure Compression

The maximum compressive stress level is a unitless ration between the maximum compressive stress experienced by the concrete ($\sigma_{c,max}$) and the fatigue reference compressive strength ($f_{ck,fat}$) which was calculated as 75 MPa according to Eq(5.1-110) in the fib Model Code 2010.

fib Compression and Tension Fatigue with $\sigma_{ct,max} > 0.026 \mid \sigma_{c,max} \mid$

The S-N curve displayed in Figure 2.7 is defined by the following equation: $logN = 12(1 - S_{c,max}$

and is valid under the following circumstances:$\sigma_{ct,max} > 0.026 \mid \sigma_{c,max} \mid$

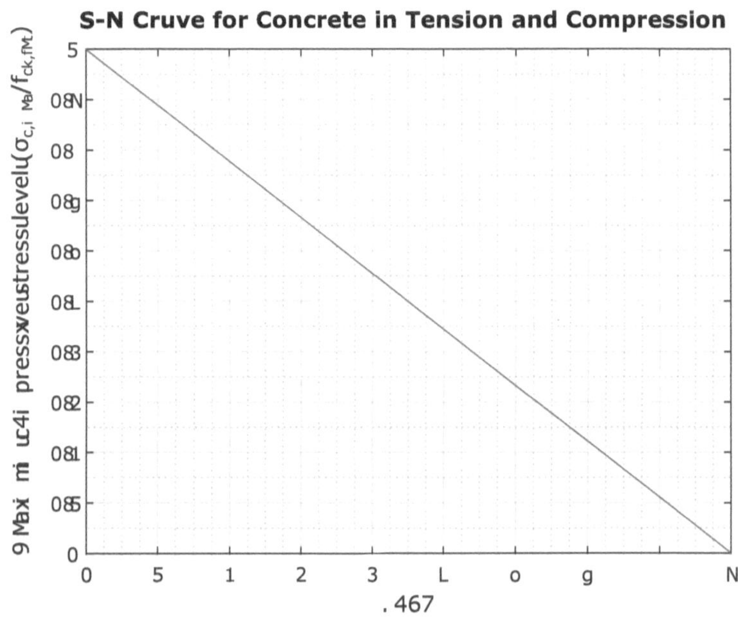

Figure 2.7: S-N curve for Concrete in Compression-tension

fib Pure tension and Compression-tension Fatigue with $\sigma_{ct,max} <= 0.026 \mid \sigma_{c,max} \mid$

The 'maximum stress level' for this graph is defined as: the maximum tensile stress experienced in the concrete (MPa) divided by the minimum characteristic tensile strength which was calculated as 76 MPa.

The S-N curve displayed in 2.8 is defined by the following equation:

$logN = 9(1 - S_{c,max}$

and is valid under the following circumstances:$\sigma_{ct,max} <= 0.026 \mid \sigma_{c,max} \mid$

41

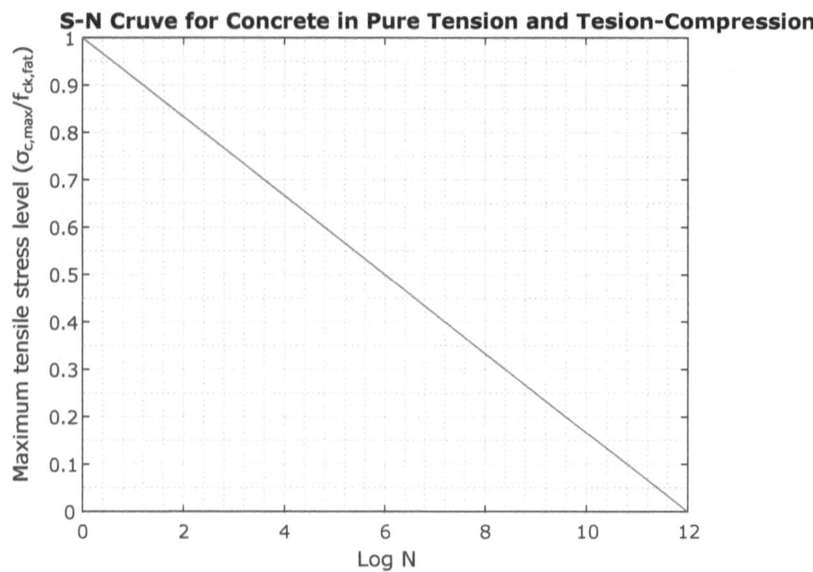

Figure 2.8: S-N curve for Concrete in Pure Compression

According to the fib Model Code 2010 the SN curve for steel consist of two segments of different slopes. Figure 2.9 illustrates the domain of these two segments.

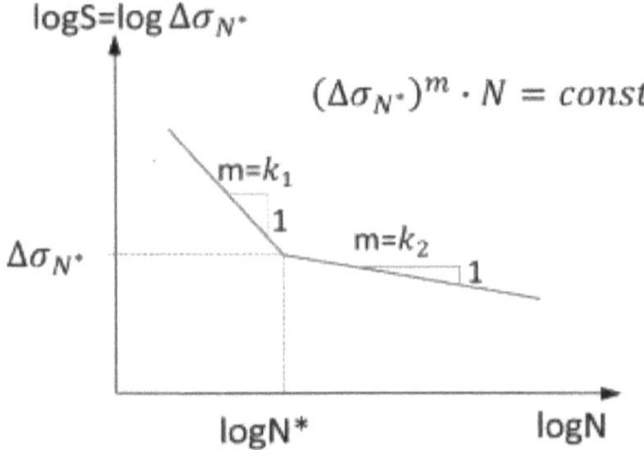

Figure 2.9: S-N curve for Prestressing Steel strand You et al. (2017)

Figure 2.10 was drawn by selecting the parameters applicable for the tendons in this study i.e. $k_1 = 5, k_2 = 9, N^* = 10^6, \Delta\sigma_{N^*}$ at N^* cycles =185 MPa

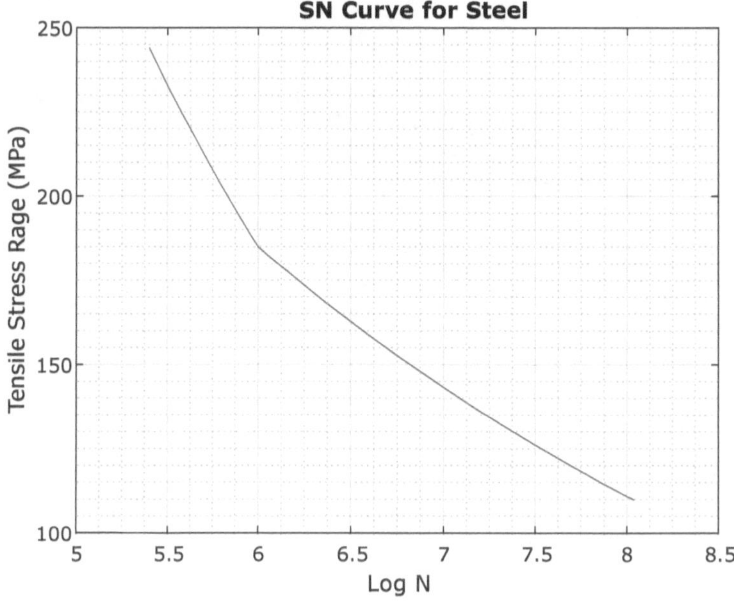

Figure 2.10: S-N Curve for 7mm Prestressing Steel Tendon

As the fib Model Code 2010 is a design guideline the S-N curves represent lower bounded data and are thus suitable for a conservative design approach. In order to achieve a more realistic representation of the steel fatigue You et al. (2017) recommends using a mean value of 300 MPa as the $\Delta\sigma_{N*}$ value rather than the 185 MPa used to draw the above graphs. This revised $\Delta\sigma_{N*}$ value will be used in the discussion portion of this report.

2.9.2 Goodman and Smith Diagrams

The Goodman and Smith diagrams are remarkably similar to S-N curves however they record the max stress on the vertical axis while the minimum stress is recorded on the horizontal axis. Different lines are drawn to indicate the number of cycles sustained before failure. Figure 2.11 illustrates a typical Goodman Diagram. see below.(Ameen and Szymanski, 2006)

44

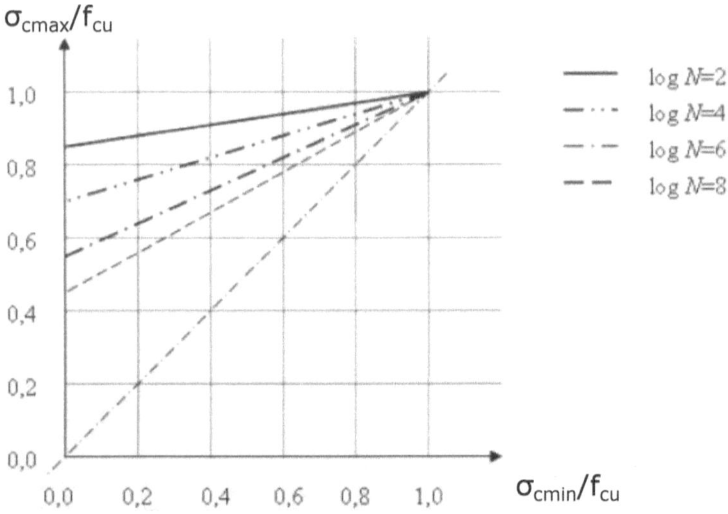

Figure 2.11: Typical Goodman and Smith Diagram for Concrete (Ameen and Szymanski, 2006)

2.9.3 Concrete Fatigue Damage Theories

Loading histories are very seldom of a fixed stress range, as assumed by the S-N and Goodman model, thus a more advanced fatigue theory was developed. Fatigue damage theories allow a wider range of loading characteristics such as varying the number and amplitude of the load, the loading sequence is not generally accounted for as fatigue damage theories assume the damage from a stress cycle is linearly additive.(Ameen and Szymanski, 2006)

Palmgren-Miner hypothesis

The Palmgren-Miner hypothesis is a commonly used linear damage model. Damage from each loading cycle is linearly summed and failure is assumed to occur when:

$$\Sigma \frac{ni}{Ni} = C \tag{1}$$

Where ni is the number of cycles at a certain load level signified by Li, Ni represents the life at load level Li. C is usually set to 1.

Some limitations of the Parmer-Miner method are as follows:

1. Due to the fact that it is a liner summation method the load history is not accounted for thus a load at the beginning or at the end is assumed to impose the same amount of damage.You et al. (2017)

2. Experimental results indicate that different damages occur when a high load is followed by a low load or vice versaYou et al. (2017)

3. The damage due to low load levels is underestimated (Ameen and Szymanski, 2006)

The damage accumulation method is an extension of the Palmgren-Miner as is used in the European and British codes (EC2,2005; BS5400, 1980). In this method the cumulative damage caused by a loading schedule with a range of different stress ranges is equated to an accumulation of equivalent fatigue stress each with constant amplitudes.You et al. (2017)

2.9.4   Constitutive models of fatigue concrete

Three distinct classes of relationships are required to describe a problem in solid mechanics; "Newtonian equations of motion, geometry of the deformations and the stress-strain relation".(Ameen and Szymanski, 2006)

The general form of the Newtonian equations such as conservation of linear and angular momentum are used to relate material stress to the applied loads. The 'geometry of deformations' equations relate the material strain to the global deformation. Constitutive equations relate the strain to the stress and thus link the kinematics(deformation) to equilibrium equations(Newtonian).(Ameen and Szymanski, 2006)

The liner elastic theory of materials is an example of a constitute equation however fatigue modes require the use of more complex relationships. The choice of the constitute equation is important so as to select an equation that best models the stress-strain relationship for a given material.

High cycle fatigue in concrete is best described by a 'damage coupled to elastic deformation' equation while low cycle fatigue is best described by 'damage coupled to plastic deformation' equation(Ameen and Szymanski, 2006).

Using the assumption that concrete fatigue dammage can be related to a reduction in stiffness, Haar and Marx (2016) conducted laboratory tests on small scale concrete specimens and used ultrasonic pulse velocity to measure the corresponding elastic modulus degradation. Haar and Marx (2016) concluded that the decrease in secant elastic modulus followed a S-shaped curve as illustrated in Figure 2.12. The correlation between the elastic modulus and the ultrasonic pulse velocity did not fit a linear relationship.

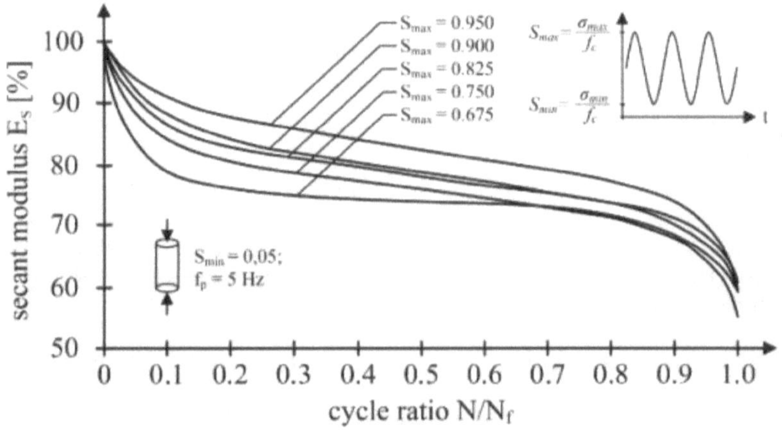

Figure 2.12: Beam 1 Load vs Displacement Haar and Marx (2016)

An example of concrete fatigue models using constitutive equations are the Modified Maekawa concrete model and Plasticity-damage bounding surface model which are best suited to describe high cycle fatigue and low cycle fatigue respectively.(Ameen and Szymanski, 2006)

2.10   Conclusion

Concrete sleepers have successfully been used to support rail tracks in many applications. Their exceptional durability makes them an attractive substitute for wooden sleepers. The primary problems associated with installing concrete sleepers on an open decked bridge are: the increased weight and their susceptibility to crack as a result of impact and fatigue loading. Minimal literature exists for the fatigue resistance of concrete sleepers. Further understanding in this area will enhance the possibility of concrete sleepers replacing wooden sleepers and thus solve a critical timber supply problem for South Africa and the world in general.

It it well established that wheel loads are a combination of quasi-dynamic forces and impact forces. Due to high intensity of the impact loading and its infrequent occurrence it could be classified as low cycle fatigue loading while quasi-dynamic loading would attribute to high cycle fatigue damage. Literature suggests that impact loading is responsible for the majority of damage to railway sleepers however there is little understanding on the extent that the high cycle quasi-dynamic fatigue plays on the damage process. Interestingly the Australian railway code (AS 1084.14-2012) specifies a fatigue test that applies a load schedule that in nature corresponds to quasi-dynamic loading. In an effort to produce results that are comparable to a recognized fatigue testing standard, this study will investigate high cycle fatigue caused by quasi-dynamic loading.

It must be noted that because neither the impact loading nor the environmental effects of fatigue are included in this study the results cannot be directly applied to predict in service fatigue life.

This **book** will study a single type of concrete sleeper (low profile sleeper) with a simple support system. The support conditions of sleeper near a bridge abutment are extremely stiff due to the close proximity to the large bridge abutment foundations. Consequently the stiff support condition in the lab would be comparable to a sleeper near a bridge abutment.

Literature suggests that there is a persistent drive to design railway components according to a damage or serviceability limit state as opposed to stress limit states. This study explores the extent to which fatigue deterioration affects the serviceability state.

FEA modes are useful in that they allow stresses to be studied for a wide range of load combinations however they need to be tuned to match the real behaviour as much as possible. This tuning will be done by comparing the displacement between the model and the lab tests. The CDP constituent equation will be used as it is flexible and open to fatigue modelling for a later study.

Fatigue will not be modelled using numerical model. Although the CDP model has the ability to be tweaked so that it could be used to predict fatigue, this is a new development and would thus require laboratory fatigue results to asses its accuracy. S-N curves provided by fib Model Code 2010 will be used to provide a means of quantifying the fatigue life.

3 Methodology

The study employed both physical laboratory tests and numerical modelling. Three Low Profile sleepers were tested to failure in a 4 point flexural test. The physical test results were used to calibrate the numeric model. Two sleepers were subject to fatigue tests.

3.0.1 Four Point Laboratory Test

Figure 3.1 illustrates the four point load set up. This loading set up was chosen as it is an idealized representation of the loading of a sleeper on an open decked bridge. The four point load also creates an even bending moment between the loading points. As a result of a large portion of the sleeper experiencing the maximum bending moment the results are less prone to local material irregularities than a three point test. The actuator load is equally split and applied to the rail seat locations using a load splitter beam. The loading rate was maintained at less than 25kN/min.

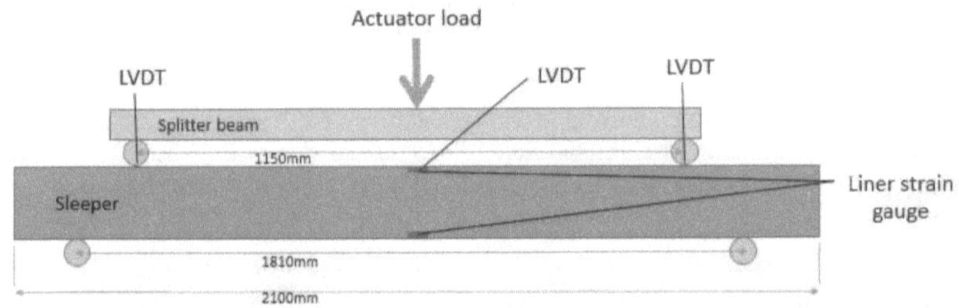

Figure 3.1: Schematic of apparatus set-up for both static tests and fatigue test

The sleepers were prepared and tested as follows:

1. The location of the loading points, supports and the measurement points was measured an marked.

2. Two strain gauges were stuck onto the sleeper, one one the top of the sleeper and one on the bottom

3. The sleeper was placed on the supports using a hoist and the loading beam is placed on top of the sleeper

4. The strain gauges and LVDTs were checked to see if they were recording data

5. Loads were applied and the test proceeded

Figure 3.2 illustrates a three point load test specified by the Australina Railway code AS 1084.14-2012. In order to produce comparable results to the recognized Australian standard, a distance of 330 mm from the loading point to the pin support was maintained in the four point bending test. Maintaining this distance ensued that the same bending moment as would occur in my this study as would occur in the standardized test.

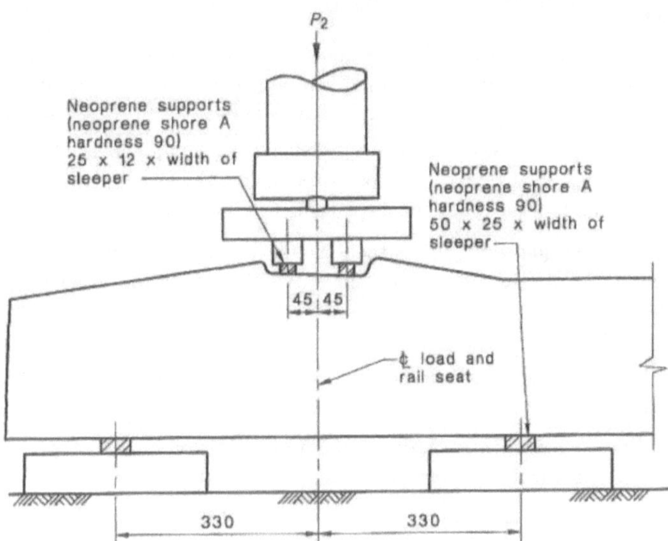

Figure 3.2: Rail seat positive moment and fatigue test set up (AS 1084.14-2012)

3.0.2 Fatigue Tests

The fatigue tests were conducted on the sleepers using the same set up as the 4 point static tests however the loading schedule was conducted according to the (AS 1084.14-2012) Australian standard. This standard is presented in Appendix D

3.1 Experimental Control and Instrumentation

Load Control vs Displacement Control

The hydraulic actuator is controlled using a proportional integral derivative (PID) control system. Either a load cell or an linear variable differential transformer (LVDT) can be used to provide input for the PID control system. If the load cell input is used the test is said to be load controlled alternatively if the LVDT input is used the test is classified as displacement control. Load control is best suited for the majority of this study as supported by the the factors noted below:

Load Control

1. The control is independent of the rig and support displacement. This is deemed important as the supporting load cell will deflect under load and the rig is slender thus it's displacement cannot be ignored.

2. Fatigue damage progression is identified by a reduction in the specimen's stiffness reduction. The specimen stiffness is easily calculated by dividing the applied load vs the resultant displacement (measured externally).

3. Standard fatigue tests such as the S-N test are load controlled as they compare the number of cycles at a fixed load vs the life time of the specimen. As this is a fatigue life investigation load control will make the results more comparable to historic data.

4. Railway loading is determined by axle weight and train speed, thus a load control test better simulates railway loading.

5. The use of load control reduces the additional forces cased by the piston's inertia which can be significant under high speed dynamic tests.

6. As precast concrete sleepers are extremely stiff it is better to use load control as measuring a large load vs a small displacement will increase the accuracy of the experiment

Displacement control is not suited for this experiment for the following reasons:

1. Displacement control is best suited for specimens of low stiffens and large displacements such as elastomer materials

2. Displacement control can be used to simulate loading environments were the displacement of a structure is specified as opposed to the load it needs to carry.

3. The use of displacement control requires that the displacement measured is that of the specimen and not that of the supports or any transfer structure in-between the specimen and the point of measurement.

3.1.1 Linear Variable Differential Transducers

Three Linear Variable Differential Transformers (LVDT) will be use to measure the displacement of the sleeper. LVDT's are renowned for offering excellent repeatability and offering high accuracy. The LVDT's offer a +- 24mm Linear Range with an accuracy of 13 microns with a confidence level of 95%. The full specification sheet for the LVDT's can be found in Appendix G.

The actuator has an inbuilt LVDT, while this measurement will be noted it does not refer to the displacement of the sample as the relative movement of the splitter beam and loading frame is also included in the actuator displacement.

3.2 Sleeper Specimen

Low Profile Concrete Sleepers are locally produced and have the same height (150mm a the rail seat) as traditional wooden sleepers. Low Profile sleepers were thus selected as the test specimen as they were deemed to posses the best ability to possibly replace wooden sleepers.

Figure 3.4 illustrates the dimensions of the Low Profile sleeper

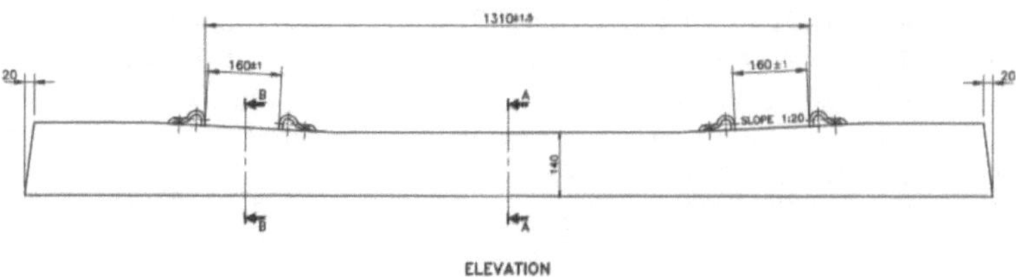

Figure 3.3: Dimensions of Low Profile Sleeper Ave (2015)

Obtaining a thorough understanding of the specimen's manufacture process and mix design is vital prior to conducting any test. The two components of the sleeper namely the concrete and the prestressed tendons will be presented below. The manufacturing process was inspected by a site visit to the Brakpan casting yard.

3.2.1 Sleeper Mix design

The concrete mix design plays an important role in determining the sleeper's characteristics during manufactures and in service.

Table 1: Sleeper Mix Design

Sleeper mix design			
	Type	**Quantity (kg)**	**Yield (liters)**
W/C	0.44		
Water		165	165
Cement	CEM I 52.5 (Afrisam)	285	90
Extender	Pozzfill	95	43
Stone	Dolerite 22mm	900	302
	Quartzite 6.7mm	250	91
Sand	Dolerite Cruser Sand	910	305
Admix	Chryso Premia 100 (superplasticiser)	2.5	2.5
Total		2607.5	1000

The following mix design is used on low profile sleepers.

The mix design is constructed to ensure the concrete meets two target strengths. The first target strength is 40 MPa at 9 hours. This rapid strength gain is achieved by the use of high temperature steam curing. The achievement of this target strength allows the sleeper to be un-moulding and prestress tendons to be released. A second target strength of 60 MPa measured at 7 days ensures the sleeper is fit for service. The specified 7 day strength is 60 MPa thus at a later age the strength will be higher than that. The fib Model 2010 code was used to calculate the strength increase in relation to time.

Knowledge of the mix design allows the elastic modulus of the concrete to be estimated.

There are no specified durability measures however by inspecting the mix design, the durability of the sleepers is expected to be good for the following reasons:

1. The a low water to cement ratio of 0.43 will create a low porosity concrete.

2. The use of a pozzolanic cement extender increases durability via reducing porosity and its chemical nature

3. The majority of the aggregates (Dolerite) have low alkali reactivity. This will ensure that Silical Alkali reaction does not occur.

3.3 Prestressed Tendons

As illustrated in Figure 3.4, Low Profile sleepers are reinforced with 9 pre-tensioned
tendons. Table 2 presents the physical properties of the tendons while Table 3
presents the mechanical properties of the tendons.

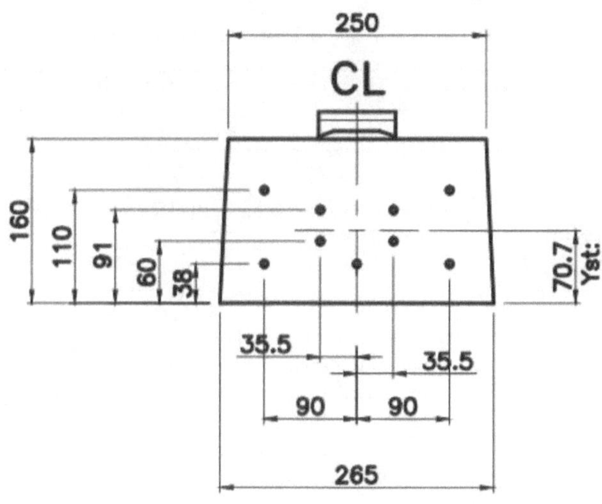

Figure 3.4: Dimensions of Low Profile Sleeper (Ave, 2015)

3.4 Finite Element Model

The commercial finite element program Abaqus was used to develop and process
the numerical model of the low profile sleeper. A concrete damage plasticity model
(CDP) was chosen to represent concrete's stress vs strain behavior while a purely

Table 2: Physical properties of Tendons

Number of tendons in sleeper	9
Diameter of tendon	7mm
Tendon surface	indented/ribbed
Effective area of steel	346.4 mm2

Table 3: Mechanical Properties of Tendon

Ultimate Tensile strength (UTS)	1670 MPa
Prestress stress	1252.5 MPa (75% of UTS)

elastic-plastic model was selected to represent the pre-stressing steel tendon stress vs stain relationship.

3.4.1 Concrete- Concrete Damage Plasticity (CDP

The CDP model is a single constitutive equation that allows irreversible damage to be incorporated for both tensile concrete cracking and compressive crushing. According to this model, the two predominant forms of concrete damage are attributed to tensile cracking and compressive crushing. The fundamentals of this constitutive theory are attributed to (Lubliner et al., 1989) The CDP constitutive equation can be divided into three components, the yield function, the flow function and the hardening rule(Lubliner et al., 1989). The hardening function and the yield function are intrinsically linked and will be discussed together.

3.4.2 Yield Criterion and Hardening Rule

A variation of Mohr-Coulomb equation was chosen as the yield function as it is important that the yield function is dependent on an explicitly defined cohesion value. The hardening function defines the evolution of the yield function by varying a scalar coefficient which reduces the cohesion value.

This relationship between the hardening rule and the yield rule results in the following being true.

1. When the material is undamaged the damage coefficient will be a minimum and the cohesion will be at it's maximum

2. When the material is fully damaged the damage coefficient will be at its

maximum and the cohesion will be at its minimum (zero)

The CDP yield function for concrete is defined as:

$$F(\sigma) = \frac{1}{1-\alpha}[\sqrt{3J_2} + \alpha I_1 + \beta < \sigma_{max} > -\gamma < -\sigma_{max} >] = c = f_c 0$$

Where:

1. $\sigma_{max} >=$ absolute maximum stress

2. $\beta = (1 - \alpha)(f_{b0}/f_{c0}) - (1 + \alpha)$

3. $\alpha = \frac{f_{b0}/f_{c0} - 1}{2(f_{b0}/f_{c0}) - 1}$

4. f_{b0}/f_{c0} = equibiaxial compressive strength/uniaxial compressive strength

5. $\gamma = \frac{3(1-\rho)}{2\rho-1}$

6. $\rho = \frac{3+\gamma}{2\gamma+3}$

7. J_2= second stress invariant of the stress deviator

8. I_1= first stress invariant

Table 4: CDP parameter values

Constitute Equation	Parameter	Domain
	fb0/fc0	1.10-1.16
Yeild criterion	Kc	0.66-0.8

Table 4 displays the set of parameters relating to the yield function. The effect that each parameter has on the yield function will be presented below. The yield function was rearranged to be plotted using polar coordinates in the principal stress plane these plots are used to graphically represent the effect of the yield function parameters.

Kc

Kc represents the ratio of the second stress invariant $\sqrt{J_2}$ on the tensile meridian, q(TM), to the second stress invariant on the compressive meridian for a defined hydro static pressure in triaxial compression. Kc can theoretically take values defined by $0.5 < Kc \leq 1$. Practically Kc often has a value between 0.6 and 0.8.(Simulia, 2013)

Figure3.5 illustrates the effect of the parameter Kc on the yield surface. It can be seen that the smaller Kc value of 0.6 results in a larger opening angle as opposed to the Kc value of 0.8. The larger the opening angle more ductile the concrete as there will be more combination of stress variants contained within the cone. The deviaroic shape also changes, tending toward a curricular cone as Kc approaches unity.

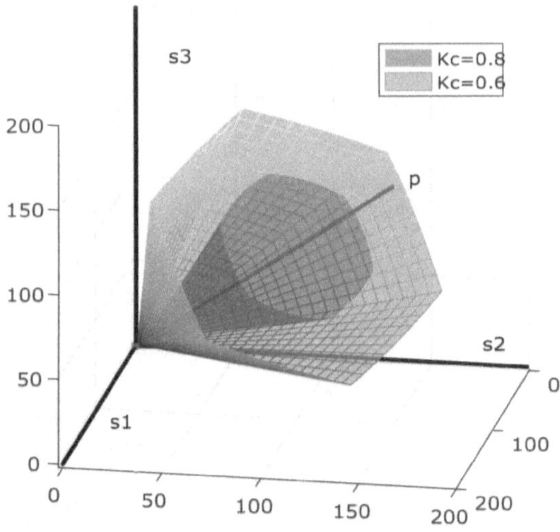

Figure 3.5: Yield envelope for different Kc values

fb0/fc0

fb0/fc0 represents the ration between the equibiaxial compressive strength and the uniaxial compressive strength. This value typical falls between 1.10 and 1.16(Simulia, 2013). As can be seen in figure 3.6 fc0/fc0 has very little influence on the shape of the yield envelope, in fact no difference can be noticed.

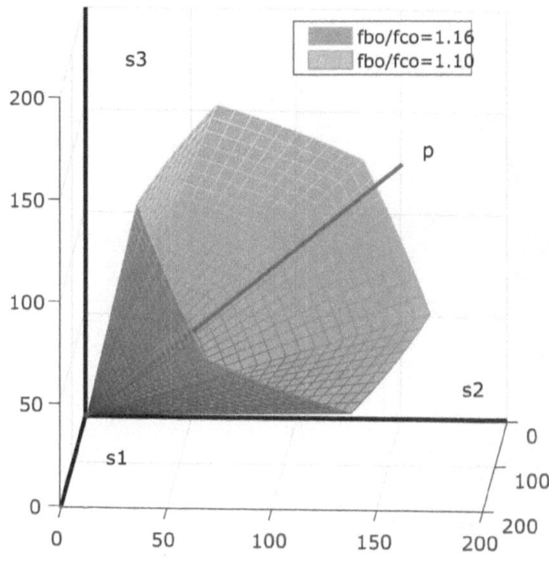

Figure 3.6: Yield envelope for different fb0/fc0 values

Uniaxial compression and tension stress vs strain graphs are used to relate the evolution of the damage variable to the evolution of the material elastic stiffness.

The assumed stress-Strain graph for concrete under uniaxial tension shown in Figure 3.7 illustrates two distinct relationships;

1. The increasing portion of the graph which is governed by initial elastic modulus (E_0) until maximum tensile stress (σ_0) is reached

2. The decreasing portion of the graph which is governed by strain softening.

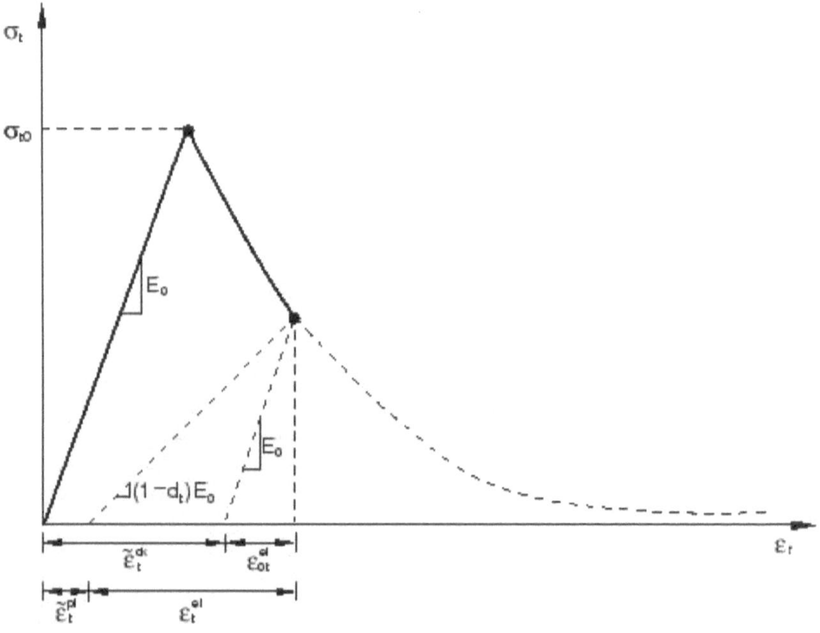

Figure 3.7: Response of concrete to uniaxial loading in tension(Simulia, 2013)

See Appendix D for the Matlab code used to draw these graphs.

Where ε_t^{pl} and ε_c^{pl} refers to the tensile and compressive equivalent plastic strains respectively. ε_t^{pl} tensile equivalent plastic strains and ε_c^{pl} compressive equivalent plastic strains are the two parameters that describe the development of the yield line.

The assumed stress-strain graph for concrete under uniaxial compression show in Figure 3.8 illustrates three separate portions.

1. The elastic portion until σ_{c0}

2. An increasing strain softening section until σ_{cu}

3. A decreasing strain softening portion

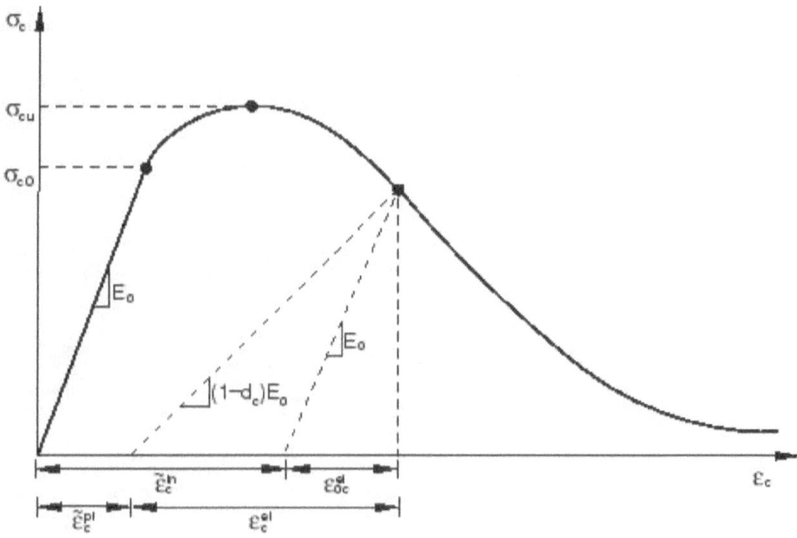

Figure 3.8: Response of concrete to uniaxial loading in compression(Simulia, 2013)

The methodology of calculating the uniaxial stress strain graphs for concrete in tension and compression and the corresponding dt and dc parameters was done according the the Lubliner/Lee/Fenves formulations as outlined by (Alfarah et al., 2017) and represented in Appendix B.

(Alfarah et al., 2017) methodology utilizes three input variables as presented below

1. Concrete crushing strength denoted by fcu

2. Element length

3. Convergence parameter

All other values such as crushing and cracking energies, elastic modulus and tensile strength are functions of these three input values. Upon inspection of how these values are calculated it is evident equations from the fib manual are used. It is possible to independently specify material properties such as cracking energies however for the preliminary numerical model (Alfarah et al., 2017)'s methodology will be used unaltered.

Table 5 gives the stress vs inelastic strain for both tension and compression as calculated using (Alfarah et al., 2017)'s methodology.

Figure 3.9 and Figure 3.10 graphically displays the values contained in Table 5.

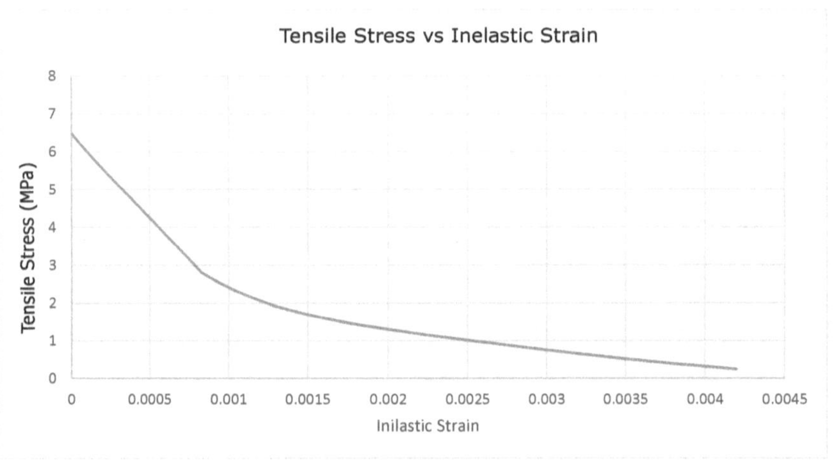

Figure 3.9: Tensile Stress vs Inelastic Strain

Table 5: Tensile and Compressive Inelastic Strain vs Strain relationship

Tensile Stress (MPa)	Inelastic Cracking Strain	Compressive Stress (MPa)	Inelastic Crushing Strain
6.5	0	32.3	0
6.4	0.0001	40.26	0.0002
2.93	0.0008	47.7	0.0004
2.42	0.001	54.75	0.0006
2.05	0.0012	61.23	0.0008
1.79	0.0014	66.88	0.001
1.59	0.0016	71.15	0.0012
1.43	0.0018	73	0.0014
1.3	0.002	71.29	0.0022
1.17	0.0022	65.07	0.0032
1.06	0.0024	56.38	0.0042
0.95	0.0026	47.31	0.0052
0.85	0.0028	39.12	0.0062
0.75	0.003	32.23	0.0072
0.65	0.0032	26.66	0.0082
0.56	0.0034	22.21	0.0092
0.47	0.0036	18.66	0.0102
0.39	0.0038	15.83	0.0112
0.31	0.004	13.56	0.0122
0.24	0.0042	11.71	0.0132
		10.2	0.0142
		7.9	0.0162
		6.29	0.0182
		5.11	0.0202
		4.23	0.0222
		3.55	0.0242
		3.03	0.0262
		2.61	0.0282
		2.27	0.0302

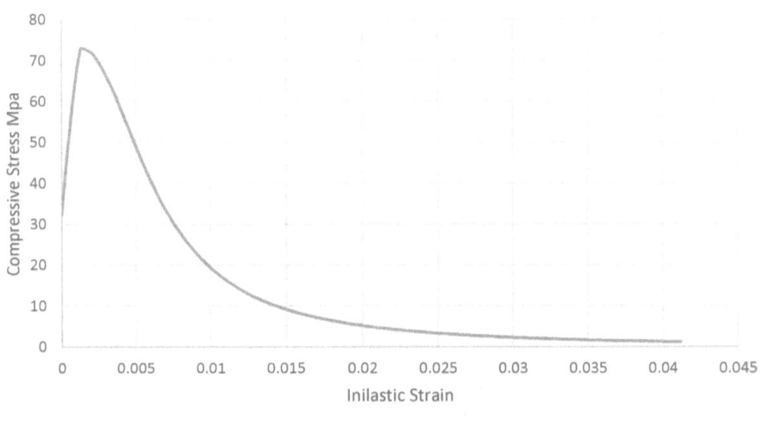

Figure 3.10: Compressive Stress vs Inelastic Strain

According to Lubliner et al. (1989) it is a justifiable assumption that cohesion is proportional to elastic stiffness and thus the hardening rule is used to predict the isotropic deterioration of the elastic stiffness of the material as plastic strain increases. This deterioration is effected by the use of a scalar damage variable acting. The damaged elastic modulus is calculated according to the equation $E = (1 - d)E_0$ Where d is a function of the d_t ,d_c and the stress state. d_t and d_c are damage variables that account for the fact that after concrete enters the strain softening portion it has a reduced or damaged elastic modulus. d_t accounts for damage due to tensile plastic strain (cracking) while d_c accounts for damage resulting from compressive plastic strain (crushing). The damage variables which are a function of the plastic strain and defined in Appendix C and exist within the domains:

$0 \leq d_t \geq 1 \; 0 \leq d_c \geq 1$

An element with a damage coefficient either d_t or d_c with value 0 is undamaged while a damage value of unity indicts a fully damaged element. The rate at which the damage coefficients increase with regards to the plastic strain provides an measure of the ductility of the material. A rapid increase in damage coefficients indicates a brittle material. Table 6 contains the values of dt and dc and their associated inelastic strain values. These values were imported into the Abaqus software. Abaqus automatically converts the stress- inelastic data into stress vs plastic strain curves.

Table 6: Tensile and Compressive Damage Coefficients

d_t	Inelastic Cracking Strain	d_c	Inelastic Crushing Strain
0	0	0	0
0.06	0.0001	0	0.0002
0.44	0.0008	0	0.0004
0.52	0.001	0	0.0006
0.6	0.0012	0	0.0008
0.66	0.0014	0	0.001
0.71	0.0016	0	0.0012
0.76	0.0018	0	0.0014
0.8	0.002	0.053	0.0022
0.83	0.0022	0.304	0.0032
0.86	0.0024	0.409	0.0042
0.88	0.0026	0.505	0.0052
0.9	0.0028	0.59	0.0062
0.92	0.003	0.662	0.0072
0.93	0.0032	0.723	0.0082
0.94	0.0034	0.774	0.0092
0.95	0.0036	0.816	0.0102
0.96	0.0038	0.851	0.0112
0.97	0.004	0.879	0.0122
		0.902	0.0132
		0.921	0.0142
		0.936	0.0152
		0.949	0.0162
		0.959	0.0172
		0.967	0.0182
		0.973	0.0192

Figure 3.11 and Figure 3.12 graphically displays the values contained in Table 6.

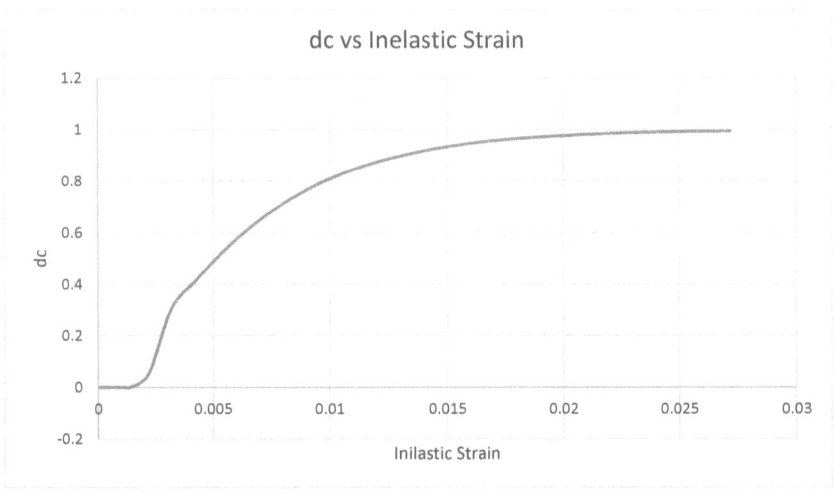

Figure 3.11: Compressive damage coefficient (dc) vs Inelastic Strain

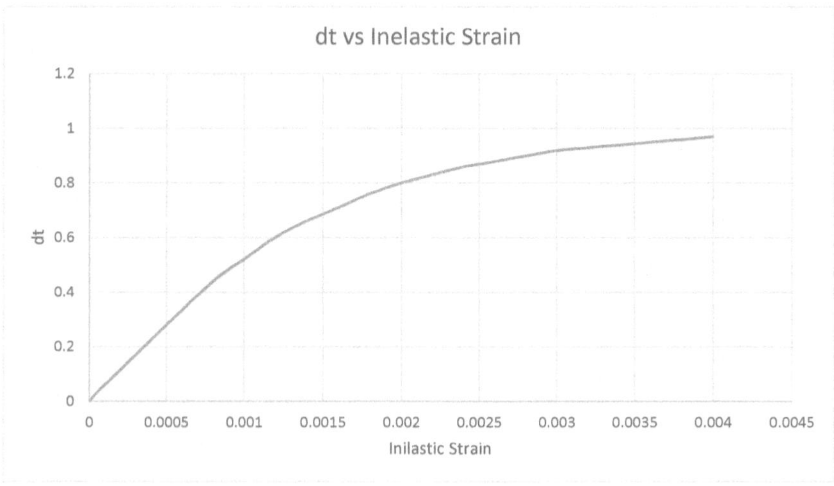

Figure 3.12: Tensile damage coefficient (dt) vs Inelastic Strain

The tension softening can be be controlled by either a post failure stress-strain relation or via the use of a fracture energy cracking criterion. The fracture energy method will be used as it is less susceptible to mesh sensitivity.

3.4.3 Flow Function

The plastic response of concrete is modeled using the Drucker-Prager hyperbolic function which in a non-assosiative flow rule in the scence that the placticity equation is not related to the yeild rule(Mor-Coulomb). The Drucker- Prager function is defined by $G = \sqrt{(e f_{etm} \tan \phi)^2 + q^2} - p \tan \phi$. Where e, f_{etm}, q, p, ϕ represents the flow potential eccentricity, the uniaxial tensile strength of the conrete,the von Mises equivalent effective stress, hydrostatic pressure and the dialation angle in the p,q $(p = \frac{1}{3} I_1 \ q = \sqrt{3 J_2}$ plane respectivly. The values of the input variables used in this calculation are similar to those presented byAlfarah et al. (2017) with modifications suggested by Szczecina and Winnicki (2015).

Dilation Angle (ϕ)

Figure 3.13 graphical presents the impact that the dilation angle plays on the flow function. The larger dilation angle of 50° results in a much stiffer flow function than a dilation angle of 30°

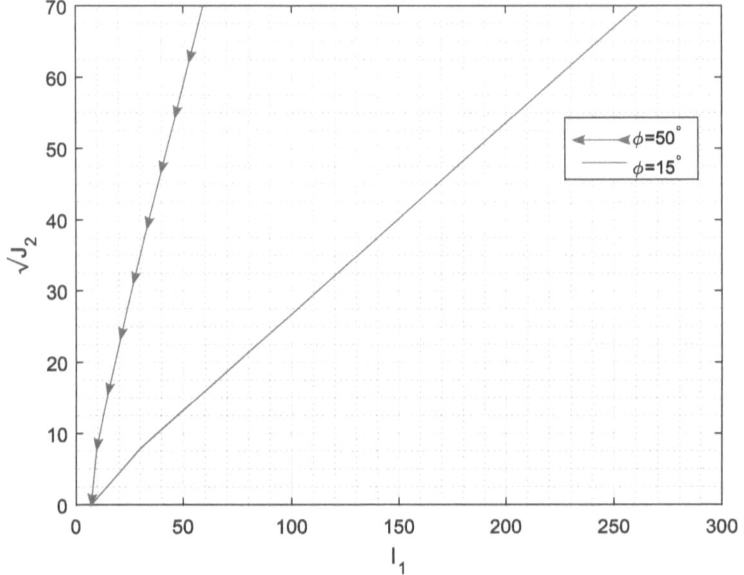

Figure 3.13: Flow Function for different values of dilation angle ϕ

The Matlab code used to draw this graph is contained in appendix C.

Eccentricity

The eccentricity in the Drucker-Prager Flow potential indicates the rate at which the function approaches the linear flow potential. A small value of eccentricity means that the materiel will have a similar dilation angle for a wider range of confining pressure while a larger value will result in the dilation angle increase for a corresponding reduction of confining stress.

3.4.4   Steel- Elastic Plastic

A simple elastic plastic constitutive equation was selected to model the steel reinforcing tendon. It was assumed that the steel would behave elastically until the yield stress was reached and then the steel would under go plastic deform. Figure 3.14 indicates the selected elastic plastic constitutive model as a stress strain graph.

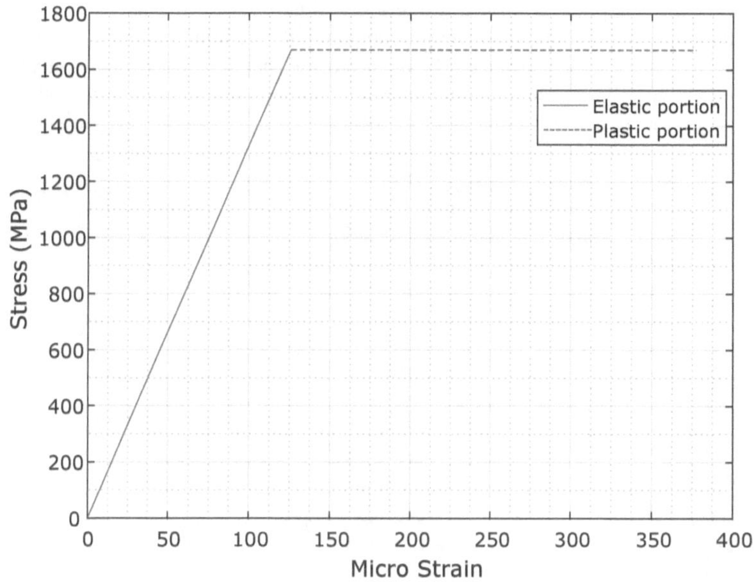

Figure 3.14: Flow Function for different values of dilation angle ϕ

The steel is assumed to be perficlty bonded to the concrete hense a recuction in bond strength or slippage can not be modeled. This simplified approach is commonly used in the analysis of precast concrete. Mousa (2016) studied the effect that the quality of bond between concrete and steel has on numerous typical test results. His tests indited that a bond length reduction of 25% in a member resulted only a slight reduction (6%) in the ultimate carrying capacity of the member. The crack width was significantly effected effected. Thus I can expect the ultimate load capacity to be accurately predicted by the model while the crack width will not be an accurate representation.

3.4.5 Mesh and Element Choice

Square block elements are recommend to model concrete so as to minimize the mesh sensitive caused by aspect rations that are not unity(Simulia, 2013). Truss elements were selected to model the reinforcing strands.

A mesh sensitivity analysis was conducted to determine the required concrete element size that yields adequate accuracy. Figure 3.15 illustrates a summary of the mesh sensitivity analysis conducted for linear element sized of 50mm,30mm, 20mm and 15mm. By comparing the displacement achieved for the 15mm and 20mm elements it can be seen that adequate convergence have been achieved. Thus a 20mm element produces acceptable results.

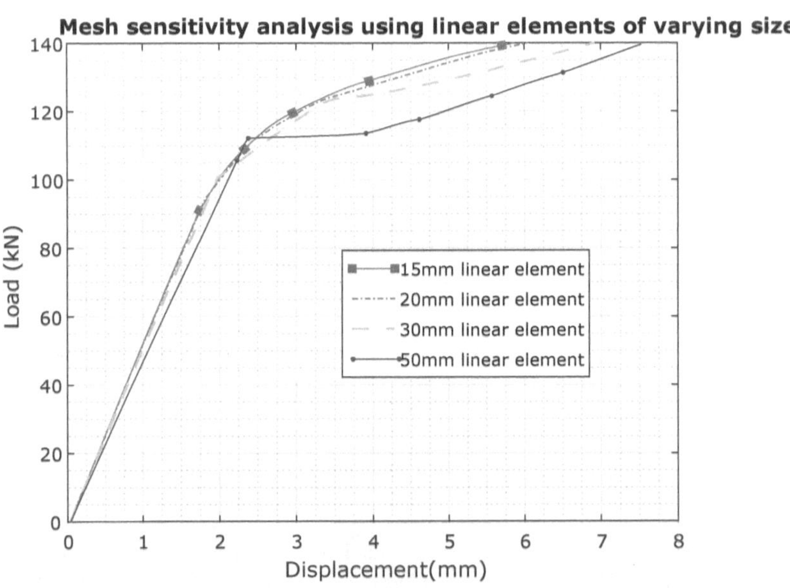

Figure 3.15: Mesh sensitivity analysis using linear elements of varying size

In an effort to reduce computational cost a quadratic element of 40mm was tested as it provides as similar accuracy to that of a 20mm linear element with a 30% reduction in computational time. Figure 3.16 illustrates the load vs displacent curve for the 15mm linear element and the 40mm quadratic element. It can be seen that the two curves a adequately similar to state that the 40mm quadratic element accurately models the beam behavior.

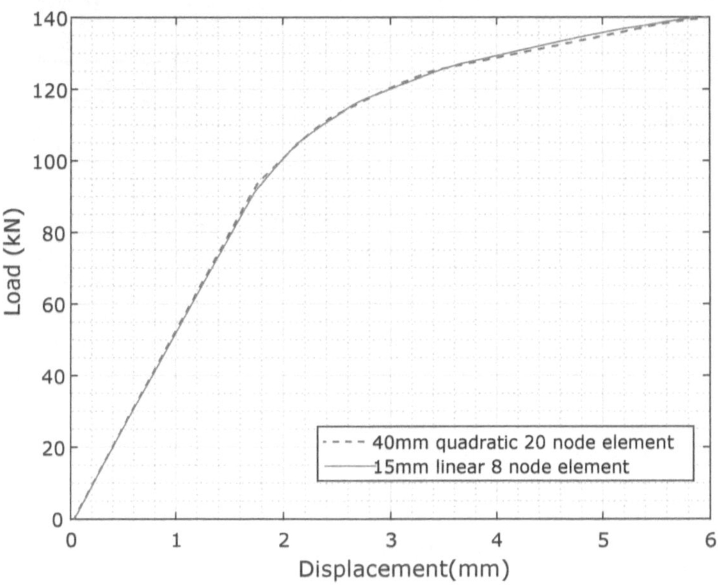

Figure 3.16: Mesh sensitivity analysis quadratic element vs linear element

Upon closer inspection it was evident that the quadratic elements were not responding correctly to the damage and crack function. Thus 17mm Incompatible mode eight-node brick element (C3D8I) was selected so as to provide adequate cracking resolution. The term 'Incompatible' refers to the fact that shear and volumetric locking, which results in artificially high bending stiffness, is removed. This numerical improvement is achieved by introducing 'bubble functions' in addition to the standard shape functions(Tanbakuei Kashani, 2017).

Table 7 contains the selected element choice and size.

Figure 3.17 and figure 3.18 illiterate the position of the nodes and integration points respectively for a C3D8I element.

Table 7: Element type and size

Material	Element name	Element size (mm)
Concrete	C3D8I: Incompatible mode eight-node brick element	17
Steel Tendon	T3D2: A 2-node linear 3-D truss	20

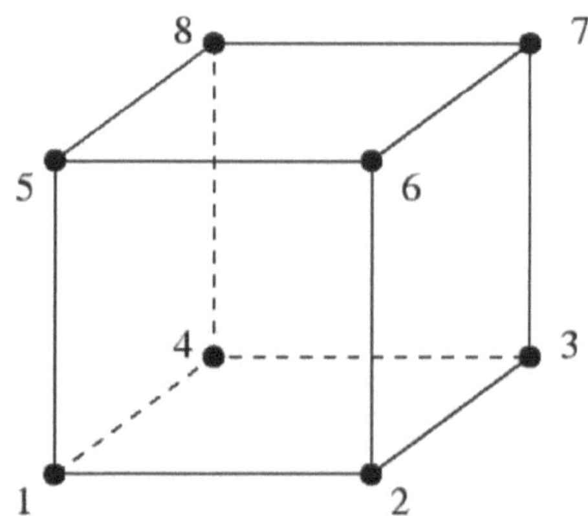

Figure 3.17: Illustration of nodes for C3D8I (c3d, 2018)

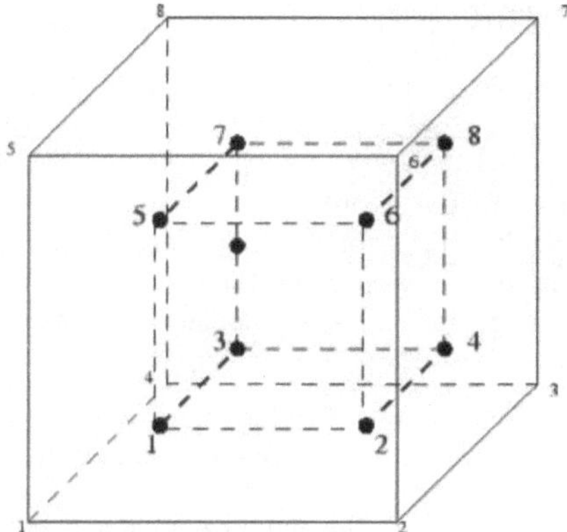

Figure 3.18: Illustration of integration points for C3D8I(Tanbakuei Kashani, 2017)

4 FEM Tuning

The concrete plasticity model has many parameters. This section will present the methodology of adjusting these parameters in order to tune the FEM model such that it best matches the laboratory test results, as measured on beam 1. The vertical displacement measured in the middle of the sleeper will be used to qualitatively compare the model results to that of the laboratory results. The strain as measured on the top and bottom of the sleeper and the crack pattern will be used to visually verify that the tuned the model exhibits a similar behaviour to that of the laboratory specimens.

Table 8 presents all parameters available for tweaking in the model. The parameters have been divided into material properties and non materiel properties.

Table 8: List of parameters used in the numerical model, five of which will be modified to tune the model of the low profile sleeper

	Constitute Equation	Paramiter	Domain	Varied or fixed
		Yeild stress of steel	1650 MPa	fixed
	Elastic	Elastic modulus of steel	210 GPa	fixed
		Elastic modulus of Concrete	40GPa-60GPa	varied
Material propeties	Flow criterion	Dilation angle	15 deg-45 deg	varied
		Eccentricy	0.1	fixed
	Yeild criterion	$fb0/fc0$	1.16	fixed
		Kc	0.66-0.8	varied
		Element length	17 mm Incompatible mode eight-node brick element	fixed
		Concrete compressive strength	40 MPa-80MPa	varied
	Assists with convergence	Viscosity parameter	0.001	fixed
Non material properties		Prestressing force	60%UTS-75%UTS	varied

The five parameters which will be varied in an attempt to tune the numerical model are identified by the 'varied' label in Table 8. The realistic domain through which these parameters can be varied is specified in the column labled 'domain'.

Steel's material properties are quoted with a 95% level by the supplier thus they were assumed to known with a very small variation and were consequently fixed thought the tuning process.

The viscosity parameter is a component of the Hillerborg visco-plastic regularization equation which is used to improve convergence during numerical computations. The model did not face any convergence problems and thus the default setting was used and fixed.

Those parameters which were found to have a small influence on the yield and flow equations will also have a minor influence on the global model results. The eccentricity and $fb0/fc0$(the ratio of the concrete biaxial compressive strength to the uniaxil compressive strength) parameters were thus fixed.

Due to the wide rage of constituent ingredients and the variation in hydration processes, concrete material properties are notoriously difficult to identify. Consequently the parameters to be tuned were those relating to the concrete material parameters and the prestressing force.

The tuning procedure involved the variation of was conducted as follows;

1. Each parameter was varied individually so as to gain an understanding of the effect that parameters had on the midpoint deflection.

2. The a combination of the parameters was selected so as to provide the closest match to the laboratory deflection curve. The 'closeness' of the match was judged visually.

4.1   Elastic modulus of concrete

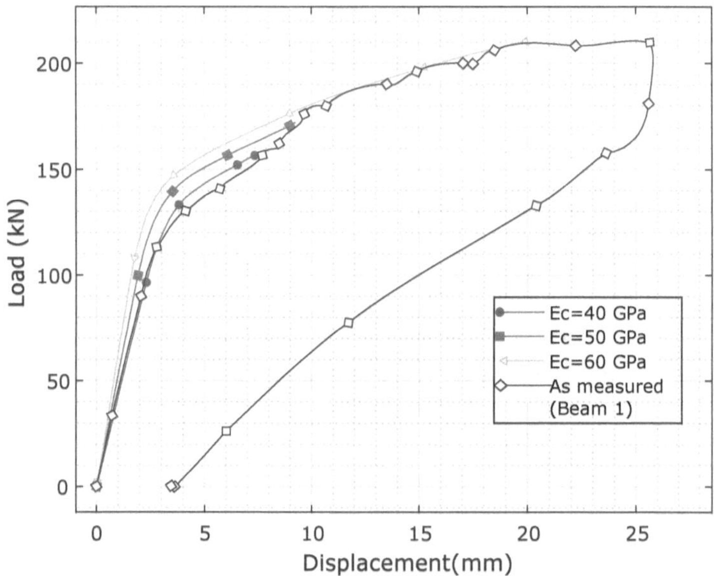

Figure 4.1: Force vs displacement relationship for different values of concrete elastic modulus

4.2 Dilation angle of concrete

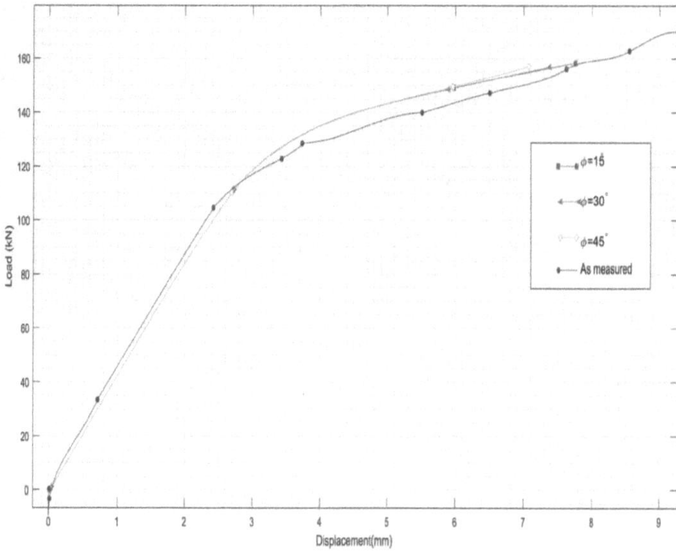

Figure 4.2: Force vs displacement relationship for different values of concrete dilation angle

4.3 Kc

Kc represents the ratio of the second stress invariant on the tensile meridian to that on the compressive meridian.

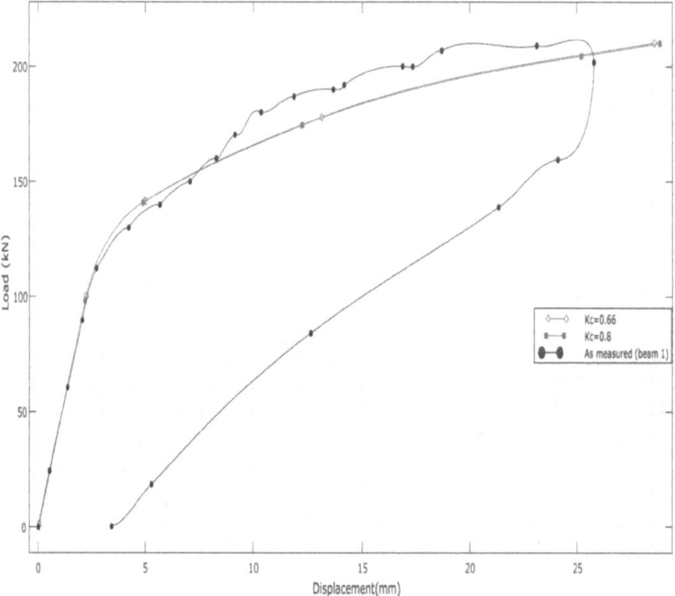

Figure 4.3: Force vs displacement relationship for different values of concrete Kc values

4.4 Concrete compressive strength

Numerous parameters are defined as a function of the concrete compressive strength and thus are varied by varying the compressive strength. The specified crushing strength for the sleepers is 60 MPa. Quality control cube crushing tests conducted by the sleeper manufacturing plant ensure that the minimum strength is met. Mean crushing strengths are typically 8 MPa higher than the specified characteristic strength and thus the actual crushing strength of the concrete in the sleepers is likely higher than 60MPa. Compressive strength of 50MPa,60MPa and 70MPa were modelled and the results are displaced below.

77

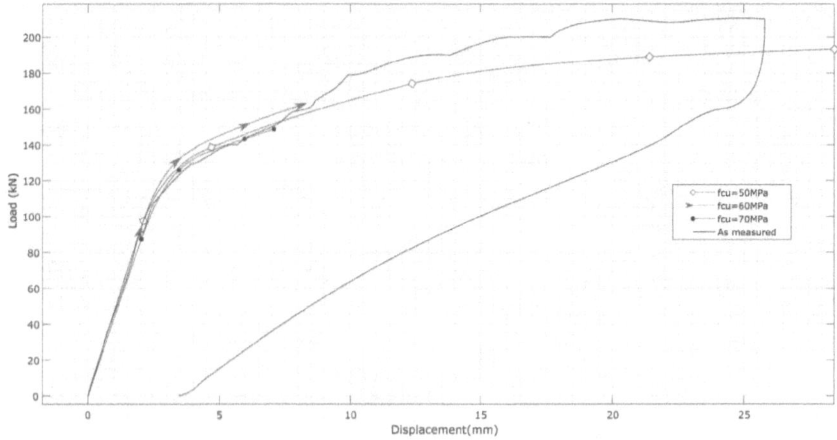

Figure 4.4: Force vs displacement relationship for different values of concrete characteristic compressive strength

4.5 Prestressing force

The design pressing force is stated as 75% of the ultimate tensile stress of the steel. It is reasonable to expect the actual prestress force to be less than that due to relaxation creep in the steel and concrete shrinkage and thus the value of 60% UTS was plotted. A much higher but unrealistic value of 80% UTS was plotted to investigate the effect it has on the model.

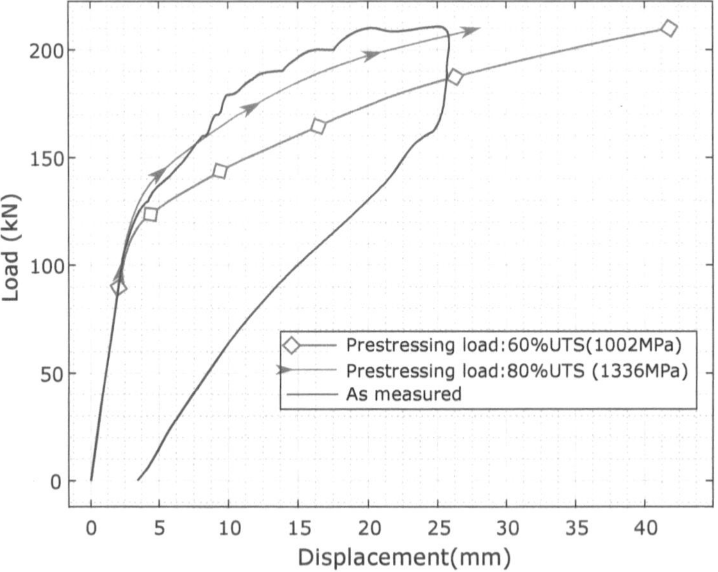

Figure 4.5: Force vs displacement relationship for different values of prestressing values

4.6 Summary of Tuning Results

The final results of the calibration process are presented in Table 9. Figure 4.6 displays the midpoint displacement vs load for the three tested beams in relation to the displacement calculate by the numerical model. The load vs displacement behaviour of the test beams can be divided into three portions, elastic, plastic and failure. It can be seen that the numerical results display a similar three part behaviour. During the elastic portion of loading (up to $\sim$100kN) the numerical result has an extremely good correlation with the test beams. During the plastic portion (between $\sim$100kN and $\sim$200kN) the numerical results has a similar curvature to that of the test beams and reached a failure point at a similar displacement value of $\sim$25mm. Failure of the numerical model was identified when the load vs disparagement curve tended to the horizontal. This indicates that there is excessive displacement with very little load capacity increase. Additional results such as crack depth/distribution and strain will be discussed in a later section.

Table 9: Selected values used in calibrated Abaqus model

	Constitute Equation	Paramiter	Selected Value
Material properties	Elastic	Yeild stress of steel	1650 Mpa
		Elastic modulus of steel	210 GPa
		Elastic modulus of Concrete	45 GPa
	Flow criterion	Dilation angle	30 degree
		Eccentricity	0.1
	Yeild criterion	fb0/fc0	1.16
		Kc	0.66
		Element length	17 mm
		Concrete compressive strength	68 MPa
	Assists with convergence	Viscosity parameter	0.001
Non material properties		Prestressing force	75%UTS

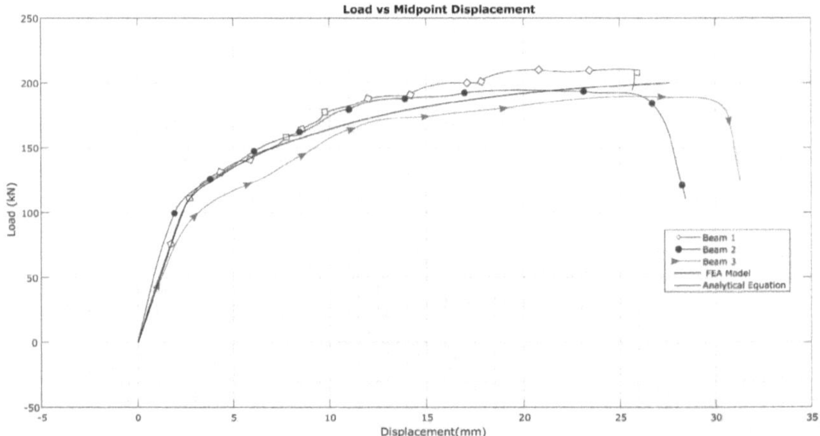

Figure 4.6: Load vs midpoint displacement

5 Results and Discussion

5.1 Introduction

The results of both the laboratory tests and the tuned FEM will be discussed and used to draw conclusions about the structural behaviour of low profile concrete sleepers. Analytical equations will be used to provide a basic understanding of the beam.

5.2 Analytical Equations

The standard beam bending equation provides a good foundation to help develop an understanding of the precast concrete sleeper's elastic behaviour.

Equation 2 represents the linear beam bending equation that relates the stress σ at a distance y from the neutral axis to the applied bending moment M and the beams second moment of area I.

$$\sigma = \frac{My}{I} \tag{2}$$

Due to the presence of prestressing tendons the standard bending equation can be modified slightly as shown in Equation 3

$$\sigma = \frac{(M - Pe)y}{I} - \frac{P}{A_t} \tag{3}$$

Where P is the total pressing force, A_t is the transformed cross section area and e is the eccentricity of the prestressing force to the neutral axis.

Tensile stresses are taken as positive and compressive stresses are taken as negative.

Using these equations some important values were calculated and are presented in Figure 5.1. The full set of calculations are displayed in Apendix A

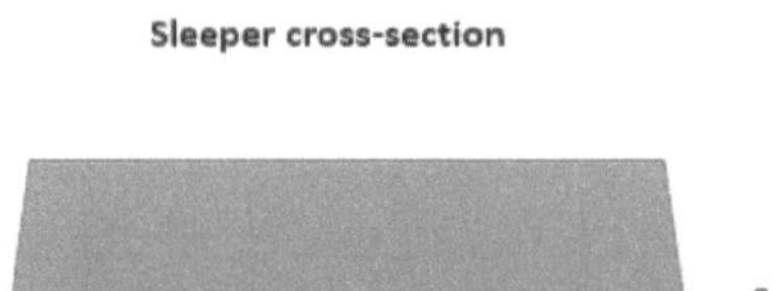

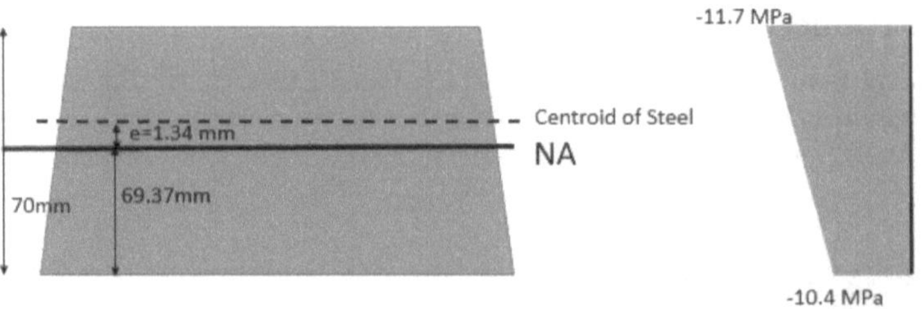

Figure 5.1: Location of neutral axis and the effects of prestressing.

The displacement of the beam's midpoint was also calculated using Equation 4 and is displayed in Figure 5.12 where it is compared both the FEM prediction and the physical laboratory rest results.

$$\Delta_{midpoint} = \frac{PL^2 a}{24EI}(3 - 4\alpha^2) \tag{4}$$

Where P is the applied load, l is the distance between the the supports, I_{trans} is the beams transformed second moment of inertia,E is the elastic modulus of the concrete, a is the distance from the support to the loading points and α is $\frac{a}{l}$

The load at which the concrete will start cracking is important since many design codes require that the concrete does not crack. This load is also referenced by the Australian standard (AS 1084.14-2012), which will be commented on later in this study. The first crack load was calculated as 82 kN and is plotted on Figure **??**

5.3 Static Laboratory Tests

Figure 5.2, Figure 5.3 and Figure 5.4 show the displacement results of a static test until failure for beam 1,2 and 3 respectively.

Beam 1

Of notice is that the Left and Right loading point stopped measuring displacement at around 150 kN, this was caused by the LVDT's running out of gauge length. In subsequent tests gauges with a larger range were used.

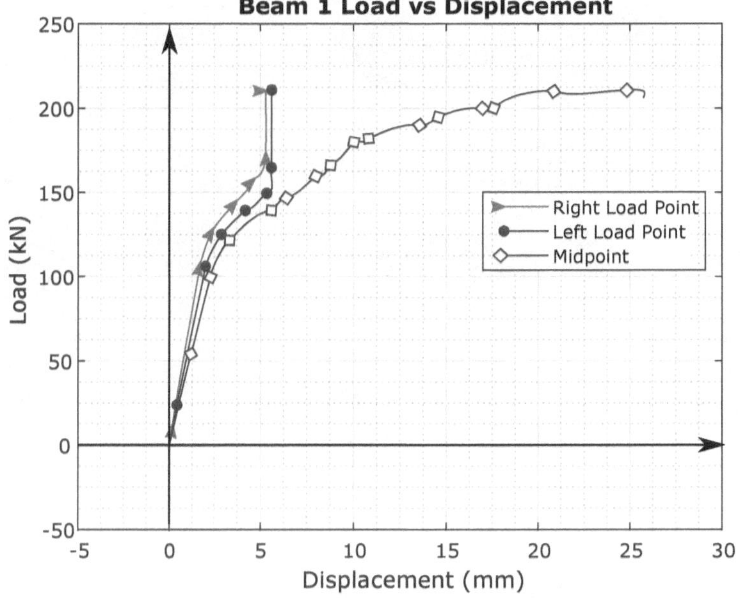

Figure 5.2: Beam 1 Load vs Displacement

Beam 2

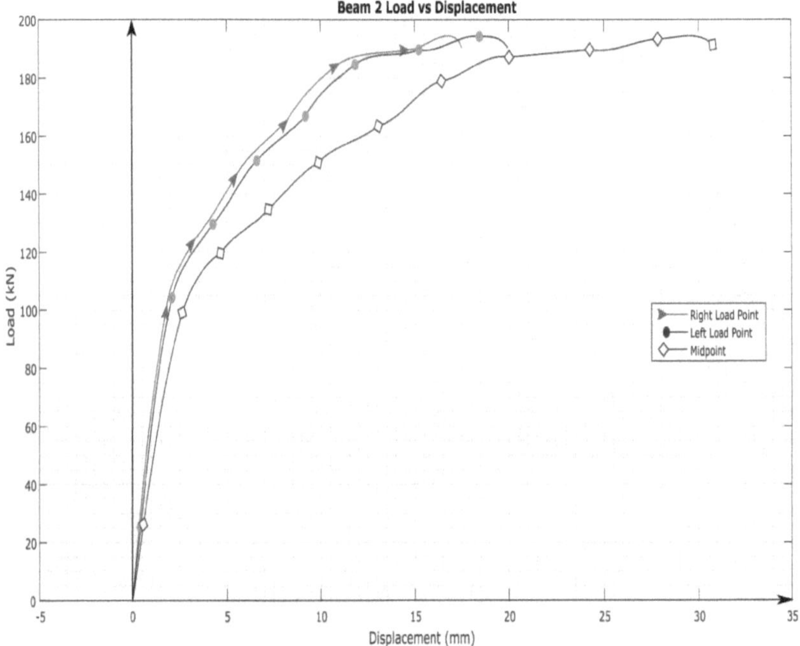

Figure 5.3: Beam 2 Load vs Displacement

Beam 3

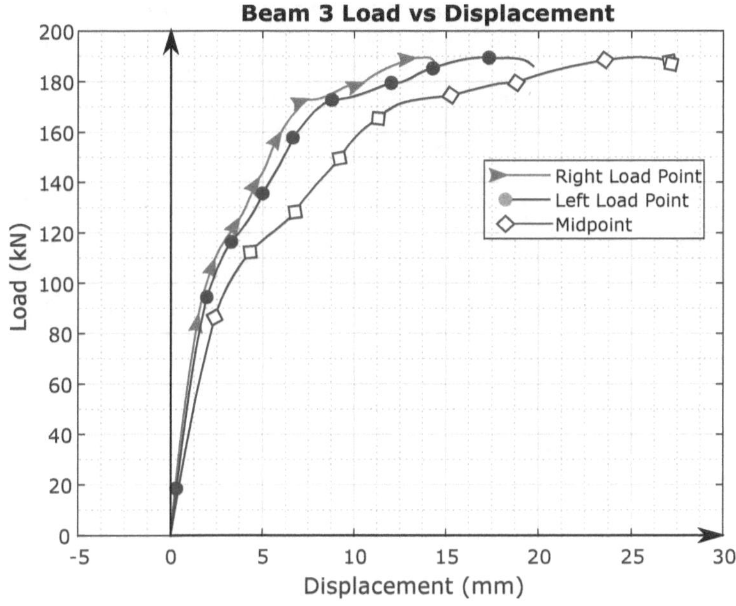

Figure 5.4: Beam 3 Load vs Displacement

As expected the displacement measured in the middle of the sleeper is large than that measured at the loading points. The displacement measured at the loading points should be identical however it can be seen for the three tests conducted that the left loading point repeatedly had a larger displacement than the right loading point. This consistent error is not due to measurement error as different LVDT's were used to measure the displacement for different tests. It is suspected that that larger displacements were measured on the left load point due to the splitter beam not eventually dividing the loads to the two load points. Only the midpoint deflection will be used for the the rest of this study.

The beam's midpoint displacements as displayed in Figure 5.5, will be used to compare load response of the three different beams. The ultimate load capacity, the displacement at failure and the shape of the load-displacement curve will be discussed. The FEA Model load midpoint displacement curve has also been displayed to indicate that its behaviour matches closely with that of the physical tests.

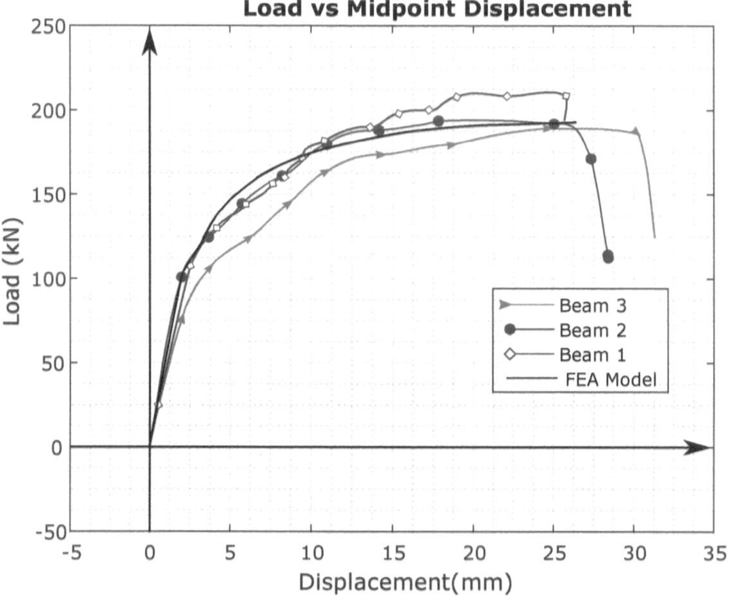

Figure 5.5: Load vs midpoint displacement for static tests

The it is evident that the load displacement relationship has three distinct phases, elastic, plastic and failure. Each will be discussed under their respective headings.

5.3.1 Elastic Portion

The elastic portion of the curve extends from initial loading until about 100kN. Thereafter the loading curve become non linear as it enters the plastic phase.

Using equation 5 and the load displacement information in the elastic portion of the curve, the elastic modulus of the concrete can be calculated theoretically.

$$E = \frac{0.5Pl^2a}{24I_{trans}}(3 - 4\alpha^2) \tag{5}$$

Where P is the applied load l is the distance between the supports, I_{trans} is the

87

beam's transformed second moment of inertia, a is the distance from the support to the loading points and α is defined as $\frac{a}{l}$

Table 10 presents the calculated elastic modulus for each of the three beams.

Table 10: Add caption

Beam	Elastic Modulus (GPa)
1	48.2
2	63.6
3	42.6
Average	**51.5**

The values in table 10 provide a domain of values from which the elastic modulus of the numerical model can be specified.

5.3.2 Plastic Portion

The plastic portion of the load displacement curve extends from 100kN until the onset of failure. The load curve asymptotically approaches the ultimate load limit. While each beam failed at slightly difference loads it is evident that all the beams failed once 27mm of midpoint displacement had been reached.

Concrete cracking is an integral component of plastic deformation and will thus be considered in more detail.

The crack depth vs applied load is displayed in Figure 5.18. Once the cracks became visible they were traced with a pen and the end of the crack was marked and labelled with the magnitude of the current load being applied. Due to the visual method of measuring the cracking progress and the delay in tracing the crack progress it is expected that the cracks occurred at a value lower than the recorded value.

Once a crack initiated it quickly extended until the level of the first reinforcing bar. For this reason the lowest recorded crack depth was 40 mm which coincides with the level of first reinforcing tendon. A line of best fit was drawn and extrapolated until it reached zero on the crack depth axis, this intercept was taken as the the load at which the first crack initiated at the sleeper soffit. An exponential function best captured the load vs crack depth relationship. A non-linear relationship is expected as the cracking process occurs in the non linear plastic region. These values are contained in Table 11.

88

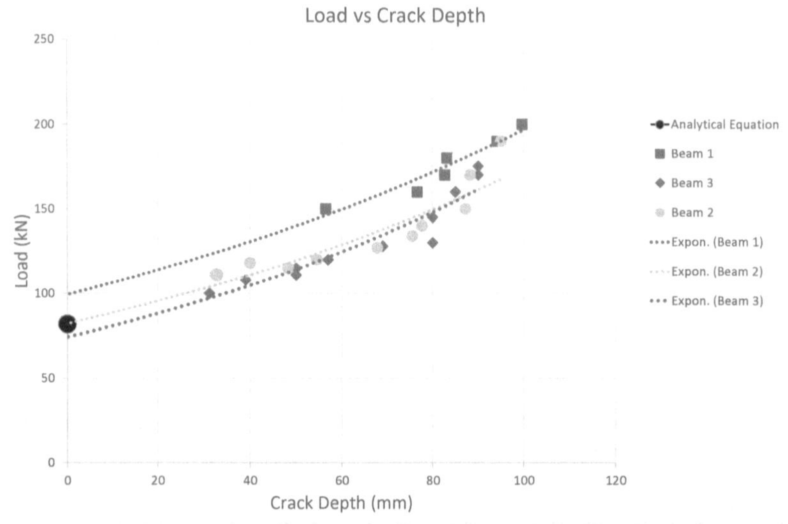

Figure 5.6: Load vs Crack Depths for static tests

Table 11: Load (extrapolated) at which first crack appears

Specimen	Load (kN)
Beam 1	99.4
Beam 2	82.3
Beam 3	74.2
Average	85.3

5.3.3 Failure

Sudden failure occurred after a period of excessive displacement while the load capacity remained constant. As was expected all specimens failed in flexure.

Table 12 presents a summary of the the ultimate load and displacement at failure. There is a 10% difference between the highest failure load and the lowest failure load.

While there is a 16% difference between the maximum and minimum displacement
at before sudden failure occurred.

Table 12: Ultimate load supported by sleepers

	Point of Failure	
	Load (kN)	Displacemnt (mm)
Beam 1	210	25.82
Beam 2	194	26.55
Beam 3	190	30.31
Average	198	27.56

At failure all beams had approximately 13 distinct flexural cracks. The crack spacing
was evenly distributed between the two loading positions. The cracks extended
to a similar depth thought the beam which indicates that the beam experienced
a constant moment in-between the loading points. Figure 5.7 illustrates the crack
spacing and depth for the three static testes.

Figure 5.8 illustrate a close up of the failure location for each of the three beams.
Failure could thus be expected to occurred anywhere in between these two points.
Interestingly the sleepers consistently failed close to the left loading position. Failure
close to the loading point could be attributed to stress concentrations caused by the
presence of rail connecting hooks however observing the load vs displacement curves
for the loading points it is evident that the left loading point repeatedly deflected
more than the right loading point. This is indicative of the load not being evenly
split between the loading points. Consistent failure close to the left loading position
is thus attributed to a slightly higher load being applied at this point.

Beam 1

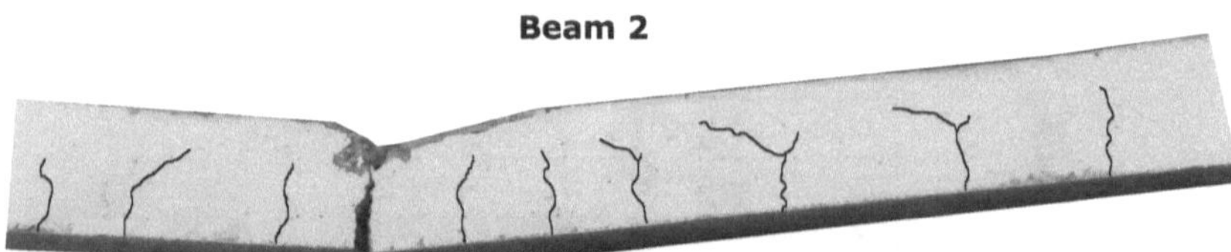

Beam 2

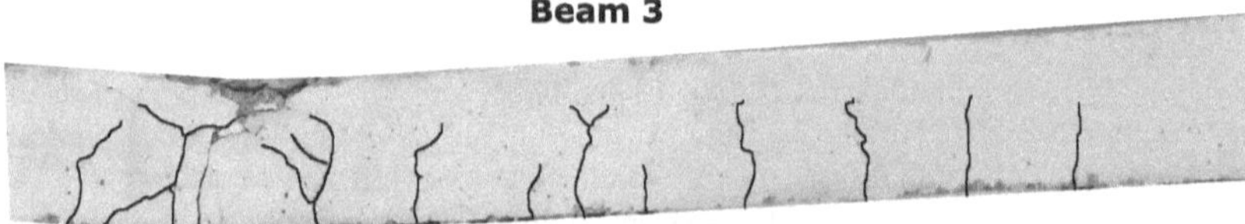

Beam 3

Abaqus Model

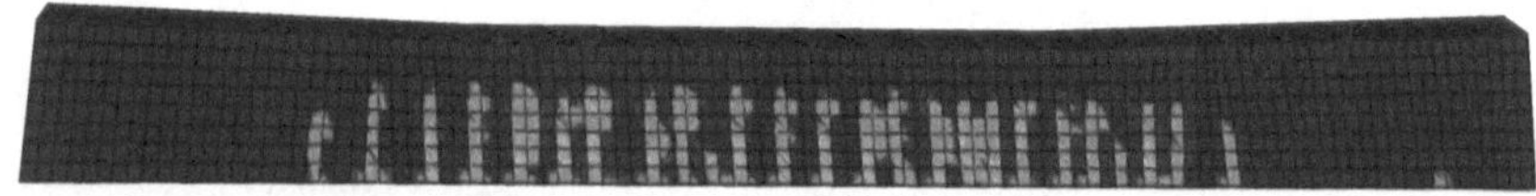

Figure 5.7: Crack spacing and depth for static beam tests

It can be seen in Figure 5.8 that significant concrete crushing occurred on beams two and three while no crushing is evident in beam one. This can be explained by understanding the mode of failure. Yielding of the steel is responsible for initial failure, once the steel has yielded any increase in load will result in significant displacement and consequently cause the concrete to crush. This is the classical failure mode of under reinforced concrete sections. During the testing of Beam 1 the steel yielded and the hydraulic actuator was stopped immediately without the applied load increasing. However during the testing of Beams 2 and 3 the hydraulic actuator continued increasing the load after the steel had yielded.

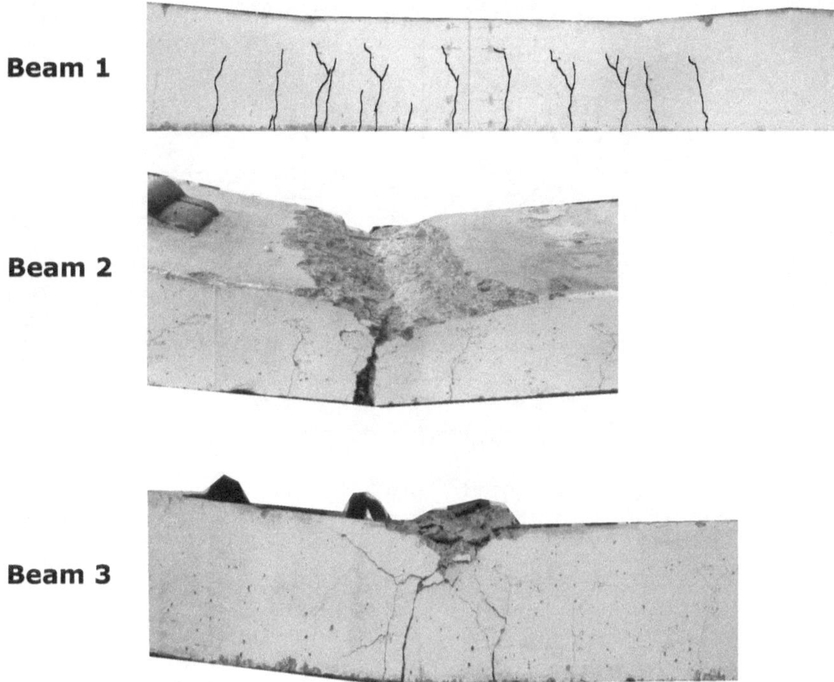

Figure 5.8: Close up of failure Location

5.3.4 Strain

Figure 5.9 shows the strain measured on to top and bottom of the three tested sleepers. The tensile and compressive strain will be discussed individually.

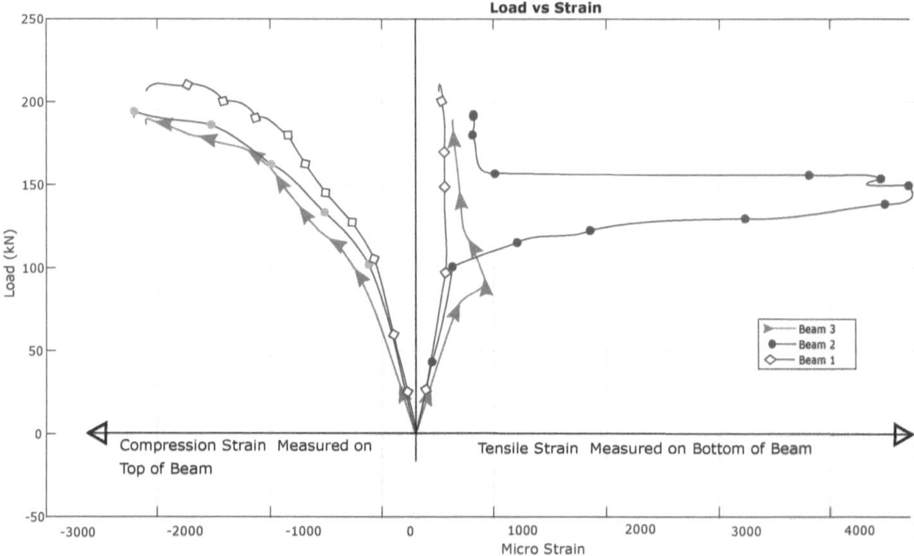

Figure 5.9: Load vs Strain

Figure 5.10 depicts the strain measured on the top surface of the sleeper the curve is continues with two clearly distinguishable segment; an elastic portion which extends until 100kN has been applied thereafter the the plastic portion extends until failure. The continues curve indicates that there was no cracking which is the expected behaviour for concrete in compression.

The stain at failure for Beam 1 and 3 was 2180 micro strain. While stain at failure for Beam 2 was and 2280 micro strain. Both these values are well below 3500 micro strain which is the generally accepted strain value at which high strength concrete under compression stress starts crushing. This observation lets us conclude that the beam failure was due to steel yielding rather than concrete crushing.

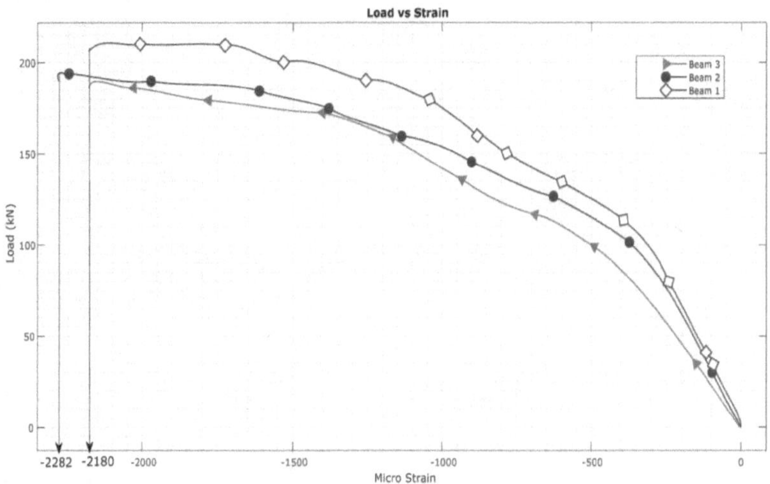

Figure 5.10: Load vs Strain

As seen in Figure 5.11 the strain measured on the bottom of the sleepers has a linear portion from zero strain until a sudden non liner increase in strain occurs. This sudden increase can be attributed to the to the onset of macro cracking occurring at a location within the strain gauge length. The strain does not continue increasing thought the test, a point exists where the strain suddenly starts reducing. This reduction in strain could either be caused a failure in the bond between the strain gauge or by the appearance of macro crack occurring on either side of the strain gauge and thus causing a stress relief in in the concrete under the strain gauge. An inspection of the gauge did not indicate any bond failure and the later reason can thus be verified.

94

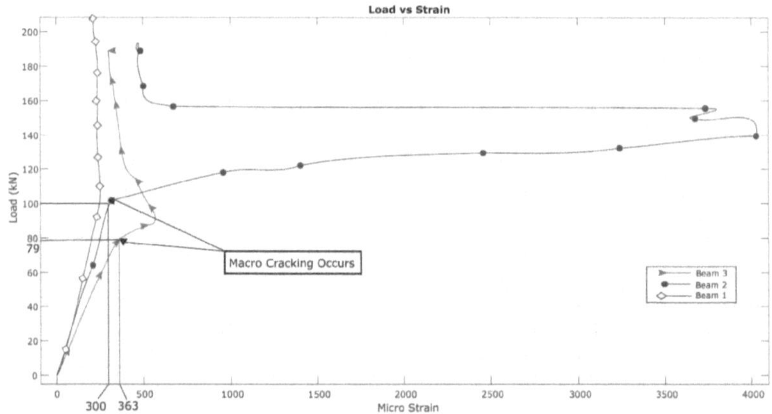

Figure 5.11: Load vs Strain Measured on Bottom of Beam

5.4 Finite Element Model results compared to laboratory and analytic results

The FEA model was tuned such that the mid point displacement closely matched that of the physical tests. A wide variety of numerically calculated results can be obtained from Abaqus. The numerical results of primary interest are the stress distribution for the concrete and steel. It is impossible to directly measure stresses experimentally and thus the tuned FEA mode is able to provide valuable information in this regard.

5.4.1 Displacement

As seen in Figure 5.12 the FEA predicted displacement matches well with the physical measurements for the elastic, plastic and failure phase. As expected, the linear analytical prediction provides accurate results until $\sim$ 100kN which is the limit of the elastic portion of the displacement curve.

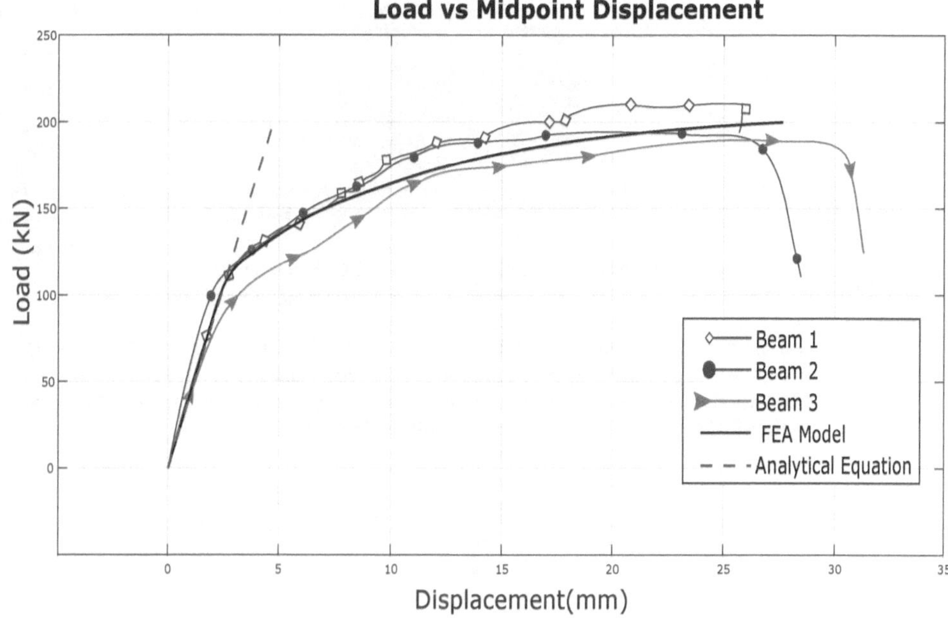

Figure 5.12: Comparison of calculated midpoint displacement and laboratory measured displacement

5.4.2 Strain

Due to the large deformations logarithmic strain was used instead of engineering strain.

Figure 5.21 shows the refernece from which the depths are measured on the sleeper.

Figure 5.13: Reference diagram for concrete stresses

It can been seen in Figure 5.14 that for the concrete above the neutral axis the compressive strain increases thought the test. The concrete, below the neutral axis, undergoing tensile stain, increases to a point 500 micro strain and then stays constant or reduces. This is as a result of strain softening occurring in surrounding concrete elements.

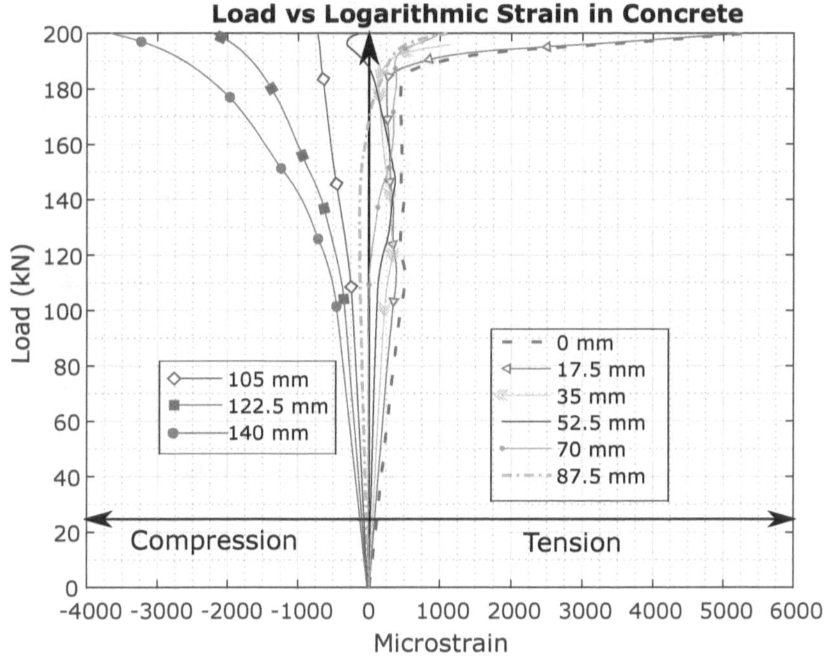

Figure 5.14: Load vs Logarithmic Strain

As seen in Figure 5.15 the FEM provides a good correlation to the physically measured strain especially during the elastic portion (up to 100kN). This indicates that the correct elastic modulus value was used in the model.

During the plastic phase of loading the FEM tensile strain suddenly increases just before failure. This can can be attributed to the the damage coefficients taking effect and simulating a of a crack opening.

The compressive strain predicted by the FEA model is consistently larger than the physical tests. At failure the compressive stress is ~ 1000 micro strain larger than the physical counterparts. This increase cannot be attributed to the numerical concrete compressive damage parameter a the crushing damage parameter was close to zero at failure.

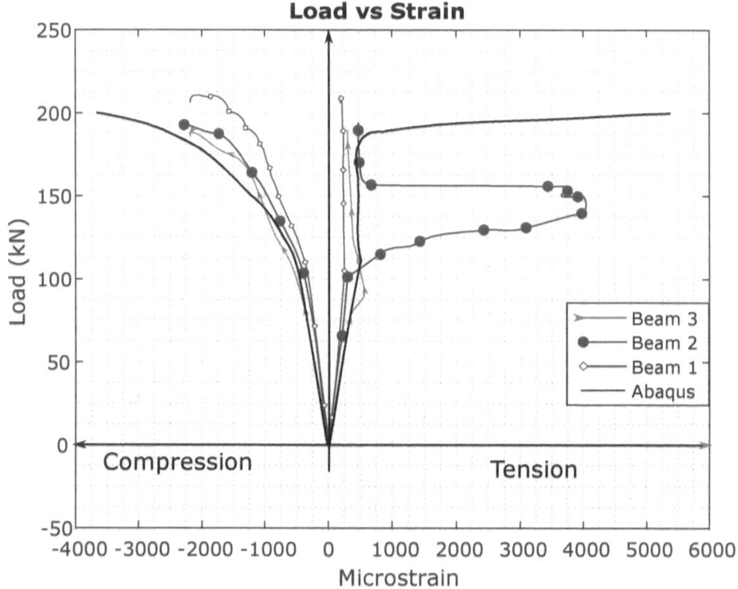

Figure 5.15: Comparison of load vs concrete strain for static tests and the numerical model

5.4.3 Cracking

Crack prediction is an important aspect in both limit state analysis and serviceability state analysis. The limit state analysis would define the occurrence of any crack as a failure while the serviceability state would provide a maximum allowable crack width before failure would be defined.

A beam's crack distribution can provide information about the mode of failure, in addition, when the crack depth is plotted against the applied loads some in-site about the beams stress distribution can be gained.

There are a number of methods available to identify cracking in the Abaqus Concrete Damage Plasticity Model. The damage coefficients dt and dc represent the deterioration of the elastic modulus of an element with regards to the strain it experiences. A damage coefficient of zero indicates an undamaged elastic modulus while a coefficient

of unity indicates a fully damaged element with zero elastic modulus. The progress
of the crack can thus be monitored via the tensile damage coefficient magnitude.

Concrete crushing can be monitored by the compressive damage coefficient dc. The
dc coefficient did not rise above zero thought the numerical simulation. This is and
indication that the concert did not undergoing crushing.

Figure ?? shows the tensile damage coefficient throughout the depth of the sleeper
in relation to the applied load. The warm colours indicate a high level of damage
while the cool colours indicate undamaged concrete.

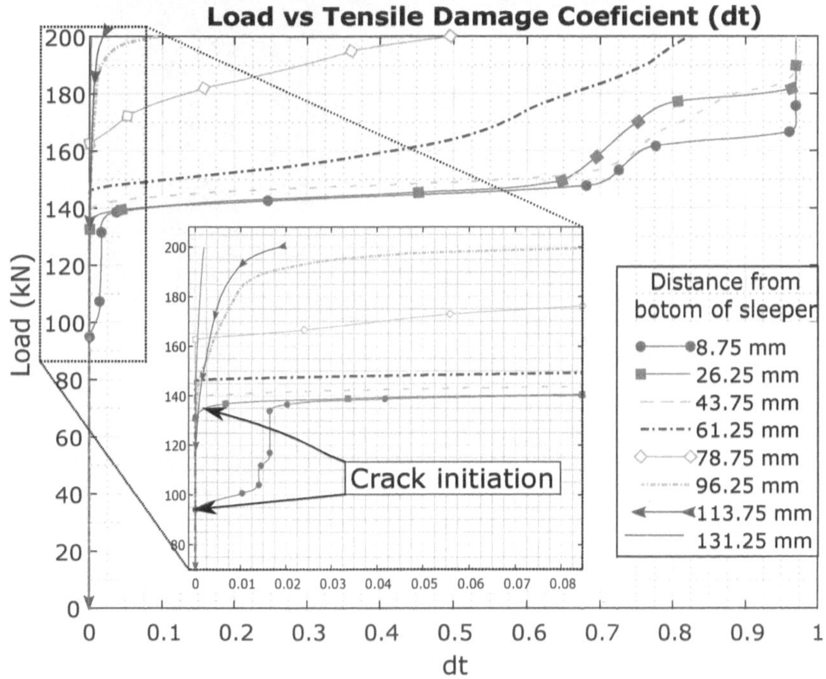

Figure 5.16: Load vs tensile damage coefficient

The close up in this graph shows the initiation of damage. It can be seen that at 8.75
mm from the bottom of the sleeper, damage started occurring at 96.5 kN applied

load. At 26.25 mm from the bottom of the sleeper, damage started occurring at 135 kN. This process of damage extending from the bottom of the sleeper to toward the top is analogous to crack propagation.

Figure 5.17 shows a colour contour of the tensile damage at a load of 200kN.

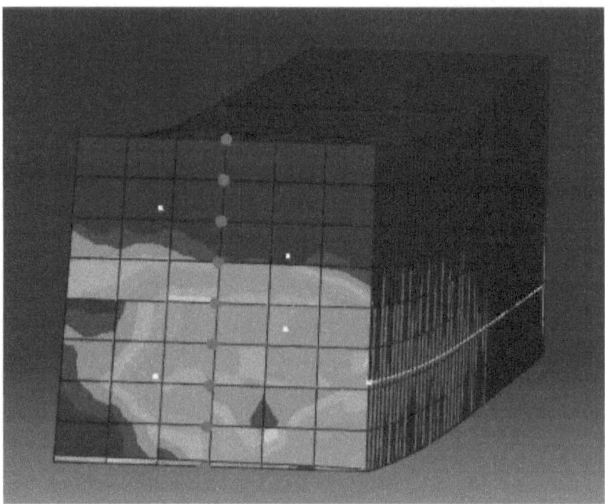

Figure 5.17: Colour contour of the concrete tensile damage coefficient at mid point of sleeper

Figure 5.18 displays the comparison between the measured crack depth, the crack depth calculated using a standard elastic beam theory and that predicted by the FEM model using the $dt > 0$ coefficient as an indication of cracking.

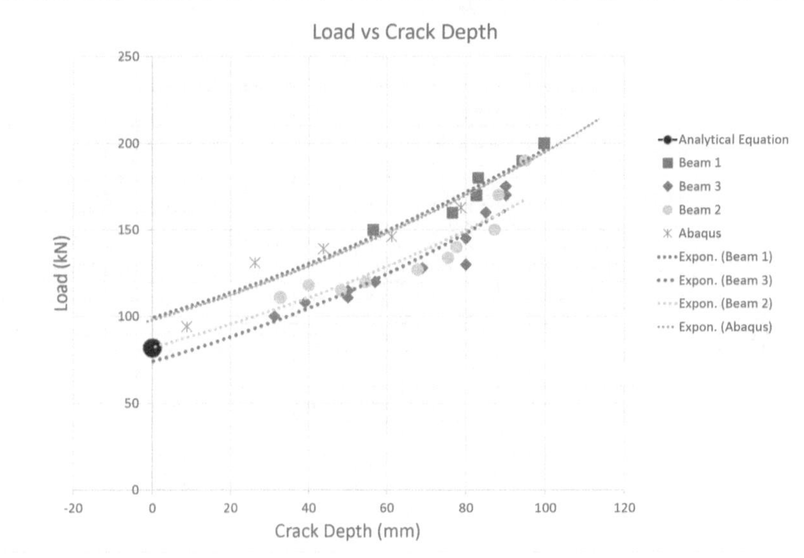

Figure 5.18: Load vs Crack Depths for static tests

Table 13 provides comparison between the measured(and extrapolated) cracking loads and those calculated by elastic beam theory and the Abaqus FEA model (also extrapolated). It must be noted that the cracking load predicted by the elastic beam theory and the Abaqus model are dependant on the assumed tensile strength of concrete. The same tensile strength of 6.5 MPa was assumed for both and was guided by the the Australin railway code (AS 1084.14-2012).

Both the elastic beam theory and Abaqus over predict the first cracking load with percentage differences of 3.1% and 14.9% respectively as compared to the measured and extrapolated crack depths. The Abaqus model could be tuned to more accurately match the crack initiation and propagation by altering the concrete material parameters such as crack energy.

5.4.4 Stress

Steel stress

The stress experienced in each of the tendons through the loading cycle is presented

Table 13: Measured and calculated cracking load

Specimen	Load (kN)
Beam 1	99.4
Beam 2	82.3
Beam 3	74.2
Average	85.3
AS-1085.14-2012	88
Abaqus	98

in figure 5.20. Figure 5.19 provides a reference for Graph 5.20. As can be seen the tendon start with a tensile stress of approximately 1000 MPa (60%UTS). This indicates that the model experiences a 15% loss in prestressing force as the steal was loaded to 75% UTS before the concrete was set.

An interesting observation is the fact that tendons in the same horizontal line i.e tendons 1 and 4 or tendons 2 and 3 have very similar starting stresses while the stresses are have as much as a 17 MPa or a 1.4% difference in there starting stresses. This phenomenon could be described by a slight lateral bending towards the what would be the axes of symmetry of the beam. The fact that the tendon closest to this line, tendon 4 has the lowest starting stress while the tendons furthest away from this line, tendons 1 and 4, have the highest starting stress. While this phenomenon would not occur if the whole sleeper were analysis the percentage error is deemed small enough to ignore.

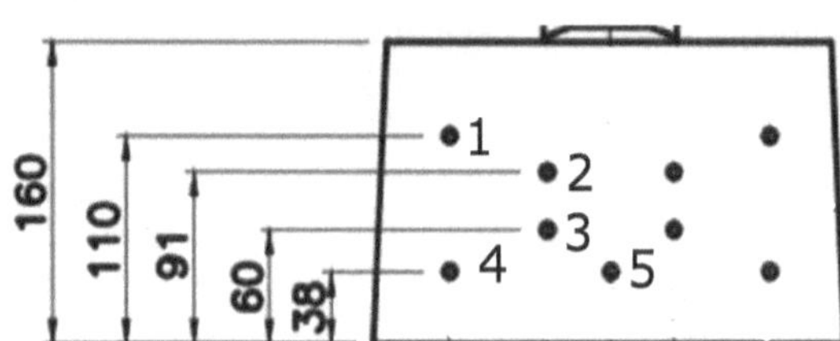

Figure 5.19: Tendon reference and location within beam

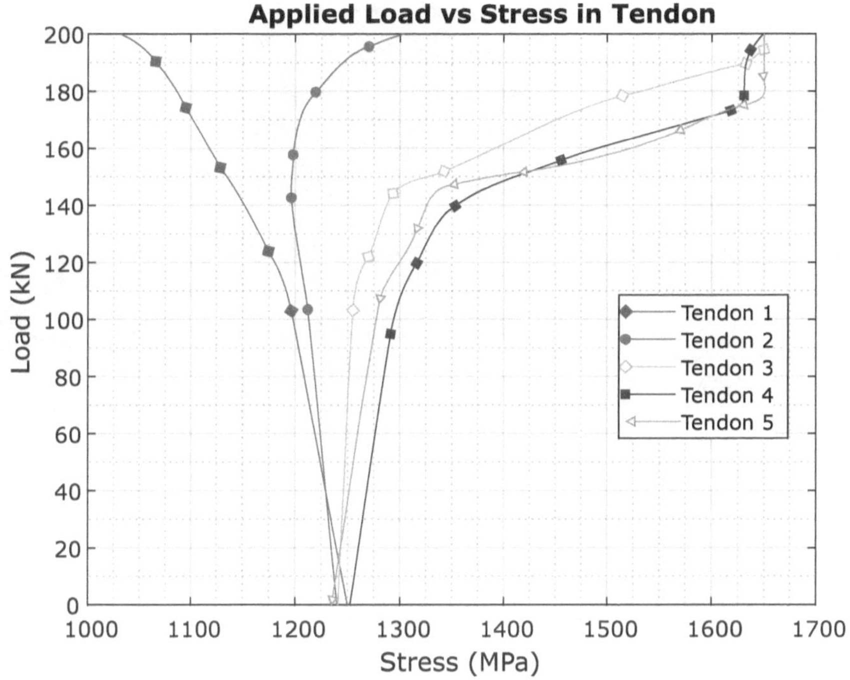

Figure 5.20: Tendon stress vs applied load as calculated by the model

Concrete stress

Figure 5.22 shows the concrete stress throughout the cross section at the midpoint of the sleeper. The stress is displayed by seven lines each of which represents the stress at a certain distance from the bottom of the sleeper.

Figure 5.21 provides the reference depths at which the concrete stresses were measured.

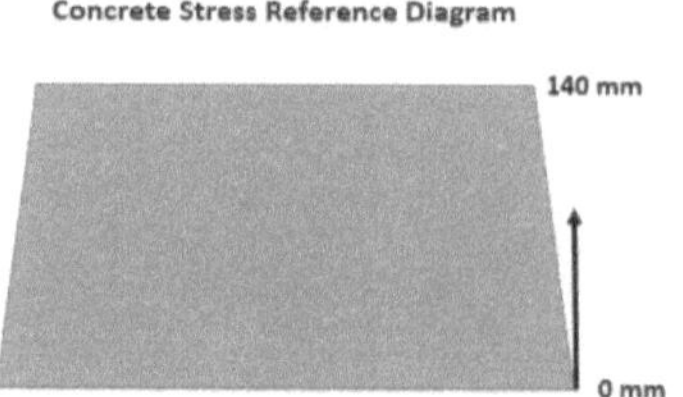

Figure 5.21: Reference diagram for concrete stresses

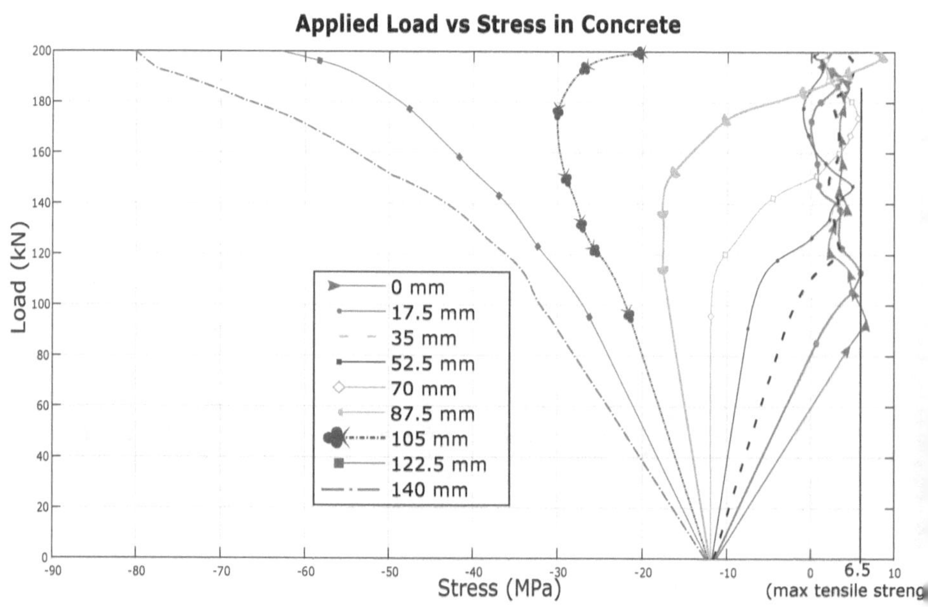

Figure 5.22: Concrete stress vs applied load as calculated by the model

The initial compressive stress due to pre-stressing will be discussed. It can be seen that at zero load the entire cross section is in compression with the stress ranging from -11.28 MPa to -12.57 MPa at the bottom and top respectively. Table 14 provides

a comparison of the concrete stresses resulting from pre-stressing for the analytic
prediction and the FEA model.

Table 14: Comparison of FEA and analytical predictions of concrete stresses

	FEA	Analytical equation	% Difference
Bottom stress (MPa)	11.28	10.4	8.5
Top stress (MPa)	12.57	11.7	7.4
Difference between top and bottom stress (MPa)	1.29	1.3	-1.0

It can be seen that the stress at a depth of 70mm does not change throughout the
elastic loading portion. This in indication that the neutral axis lies close to 70mm
from the bottom of the beam. The concrete stresses at positions below the neutral
axis decrease until they become tensile. The occurrence of cracking can be identified
when the tensile stress at that depth reaches a maximum value and then decreases
to a value close to zero.

The maximum tensile stress was recorded at the beam soffit with a magnitude of
6.6MPa at an applied load of 94.9 kN. This is an expected result as it corresponds
to the defined tensile strength of concrete and this applied load corresponds closely
to the first cracking load predicted by the damage coefficients.

As expected the location of the neutral axis moves up as the concrete starts cracking.
This is indicated by a stress reversal seen at depths of at 70 mm, 87.5 mm and 105
mm.

The maximum compressive stress experienced by the beam is indicated as 80 MPa.
This is 4 MPa higher than the specified maximum crushing stress of 76 MPa. The
fact that the compressive stress can be higher than the maximum states compressive
strength without crushing damage occurring lies in the manner Abaqus extrapolates
results from the element intergration points to the element edges. The stress at the
perimeter of an element (at the top of the sleeper) is extrapolated using a linear
shape function while The damage factor is calculated for the whole element according
the stress and strain experienced at the centroid of the element. As can be seen
in Figure 5.24 centroidal stress for one of the top most element is 70 MPa which
is less than the maximum compressive strength, consequently it is expected that
the compressive damage factor should be close to zero. The damage factor for this
element was 0.03 which indicates that crushing damage has hardly started.

Figure 5.23 shows the colour contour of the stress field present at the midoint of the sleeper under a loading of 200kN. The warm colours represent tension while the cool colours represent compression.

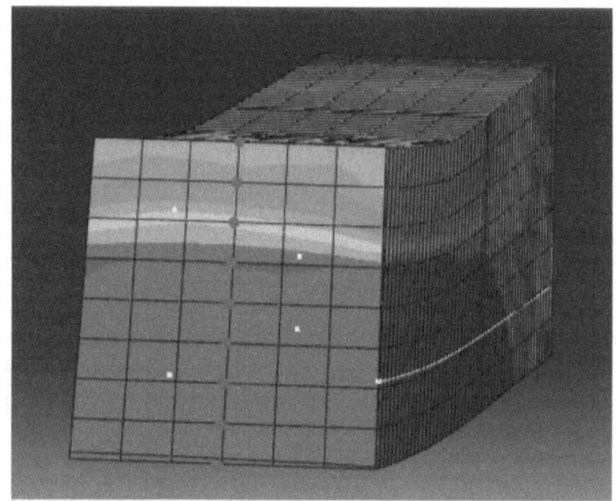

Figure 5.23: Colour contour of stress field at mid point of sleeper

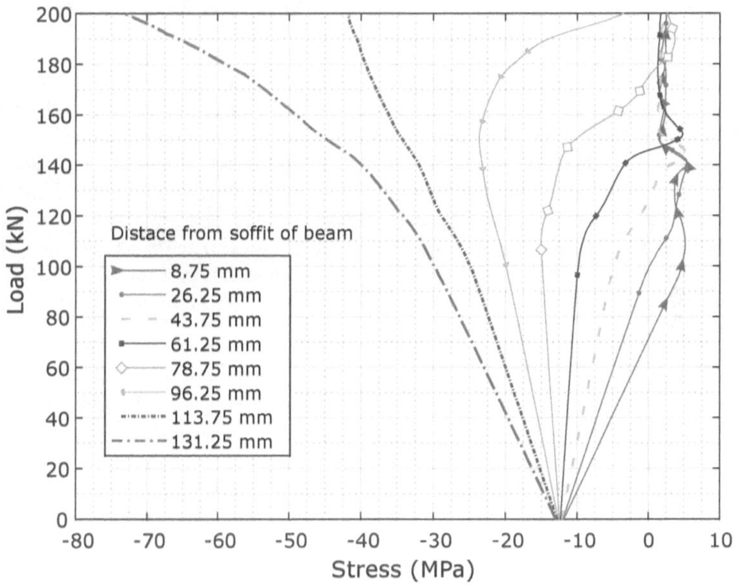

Figure 5.24: Centroidal stress of concrete vs applied load

5.5 Fatigue Calculation

Both the numerical model and the laboratory tests had monotonic loading regimes. The assumption will be made that a fluctuating load will induce the same stress field in the sleeper materials as did the monotonic load. Using this assumption, the stress output data from the FEA model and S-N curves; the fatigue life of both the steel and concrete can be predicted.

It is valuable to restate that although fatigue life is dependent on environmental factors, stress range and the load schedule (order in which loads of different magnitude are applied) only the stress range will be used to predict fatigue life.

S-N curves are developed by conducting numerous fatigue tests on specimens at a range of stress levels. It is thus important to choose an S-N curve that was developed using a material as similar to the material of which the fatigue life is to be predicted.

As presented in the literature review the fib 2010 manual contains S-N curves for prestessing steel and concrete. These curves will be used in this study and are presented under their respective headings.

As stated in the literature review, neither the model nor the laboratory accurately represents the geometry or nature of the sleeper support conditions and load application. Consequently it does not make sense to conduct the fatigue analysis using a typical rail load applied to the rail set locations, as the resulting bending moments will differ between studies set up and that found in the field. As stress is dependent on the bending moment it is important to ensure that the bending moment is related to the fatigue life rather than the load applied at the rail seats. There are many empirical equations that are used to relate the sleeper bending moment to the rail seat loads. These can be used to relate the fatigue life to the rail seat load and thus the axle load.

Transnet(the client for which the sleepers are made) uses a limit state design approach and specifies a minimum bending moment of 11.8 kNm at which the maximum compressive stress value of 3.5MPa should not be exceeded (Meyer, 2016). A load of 35.7kN is required to be applied at each rail seat in order to achieve the 11.8 kNm bending moment stated by Transnet. Consequently a 71.5 kN load should be applied to the splitter beam. It is of interest to observe what the expected fatigue life of the sleeper is at the maximum allowable stress limit.

The process by which the fatigue life is calculated is enumerated below:

1. Define the maximum and minimum bending moment for which the sleeper fatigue life will be calculated

2. Calculate the loads required to apply to the sleeper in order to achieve the desired bending moment

3. Read the minimum and maximum stress associated with the minimum and maximum loads from the load vs stress graphs produced by the finite element analysis. Use the min and maximum stress to calculate the stress range

4. Plot the stress range on the S-N curve associated with the material in question

5. Read the fatigue life from the S-N curve

5.5.1 Prestressing Steel S-N curve

The fib Model Code 2010 is a design guideline thus the S-N curves represent lower bounded data and are suitable for a conservative design approach. In order to achieve a more realistic representation of the steel fatigue You et al. (2017) recommends using a mean value of 300 MPa as the $\Delta\sigma_{N*}$ value rather than the 185 MPa which is stated in the fib Model. This revised $\Delta\sigma_{N*}$ value was used to develop the steel S-N curve which is displayed in Figure 5.25.

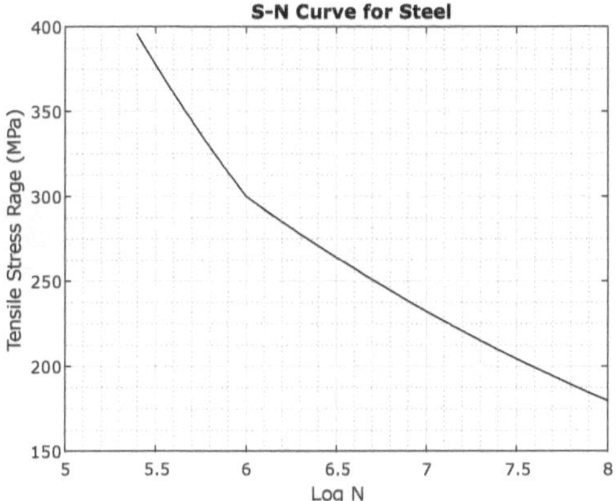

Figure 5.25: S-N curve for Steel in tension

The maximum stress range in the steel tendons resulting from a load fluctuating from zero kN to 71.5 kN is 37 MPa. This value can be obtained by using the stress vs load output of the model and is displayed in Figure 5.26. Despite the position of tendons 4 and 5 causing them to experience the highest tensile stress the stress range was greatest for tendon 1.

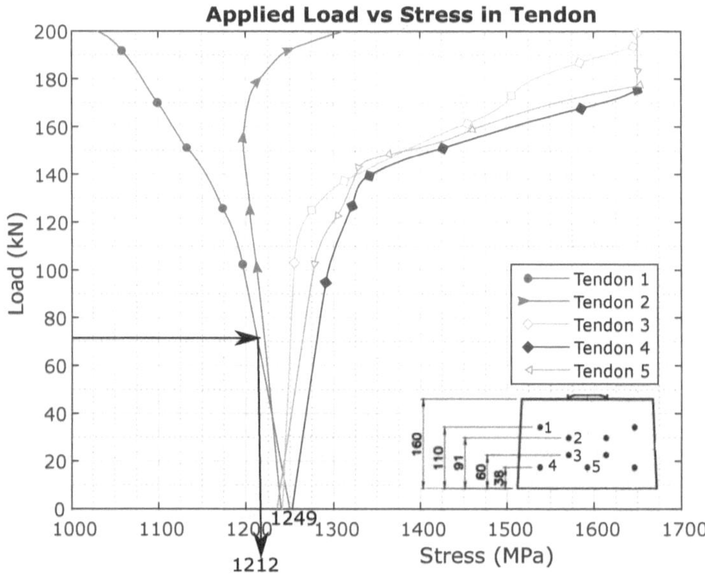

Figure 5.26: Steel stress range resulting from fluctuating load

When this stress range is plotted on the fib S-N curve, the fatigue life associated with this stress range can be calculated. It is evident that this stress range is significantly below the endurance limit of the steel and thus there will be no steel failure at his stress range. Steel exhibits a phenomenons called fatigue endurance where at or below a certain stress level the steel will not fail in fatigue.

As opposed to identifying the fatigue life for a given bending moment the inverse operation can be conducted. The applied load and thus resulting bending moment can be calculated for a given fatigue life. This will be done for the fatigue lives of one million cycles and ten million cycles. The one million cycles limit was chosen as it represents a common benchmark for a short fatigue life while the ten million cycle limit represent a long fatigue life. The stress level associated to ten million cycles is close to steel's endurance limit.

Figure 5.27 will be used to calculate the required stress ranges needed to cause fatigue failure at both one million cycles and ten million cycles.

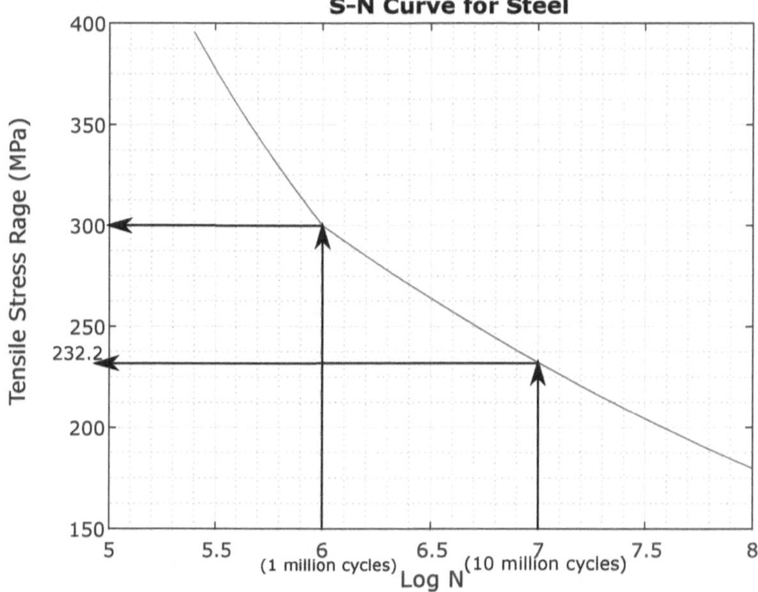

Figure 5.27: Stress level associated with one million and 10 million cycles for steel S-N curve

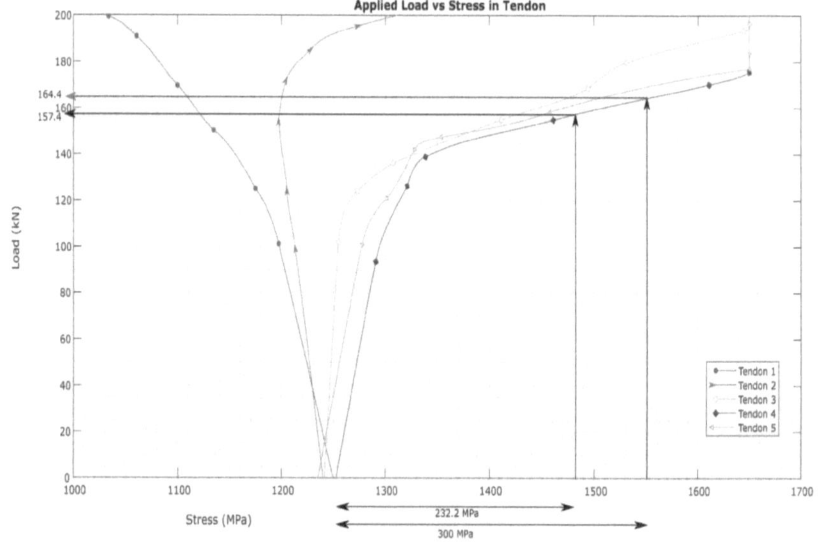

Figure 5.28: Load required to result in fatigue failure in steel tendon at one million and 10 million cycles

Table15 summarizes the required load to produce fatigue failure in the steel for one million cycles and ten million cycles these loads are read off Figure 5.28.

Table 15: Load required to produce stress levels in the steel tendons associated with a failure of one and ten million cycles

Numbe of Cycles	Stress range (MPa)	Load requied to produce failure (kN)
1 million	300	164.4
10 million	232.2	157.4

There is only a 7 kN between the load required to produce a steel failure at 1 million cycles and the load required to produce a failure at 10 million cycles. This is as a result of the sleeper being cracked at this point and the steel tendons responding in a non-linear manner. This is an interesting observation as it indicates that once a critical limit has been reached in the sleeper, fatigue failure becomes increasingly sensitive to small load increase. This sensitivity will make the fatigue failure difficult to predict.

5.5.2 Concrete S-N curve

Although current literature suggests that the steel fails in fatigue well before the concrete fails. The S-N curves used to make these claims are not presented. The fatigue response of the concrete will be investigated using a similar approach as was conducted for the steel fatigue.

Figure 5.29 is used to obtain the maximum compressive stress associated with an applied load of 71.5kN. The maximum stress associated with this load level is 24.9 MPa and will be used to calculate the fatigue life of the concrete.

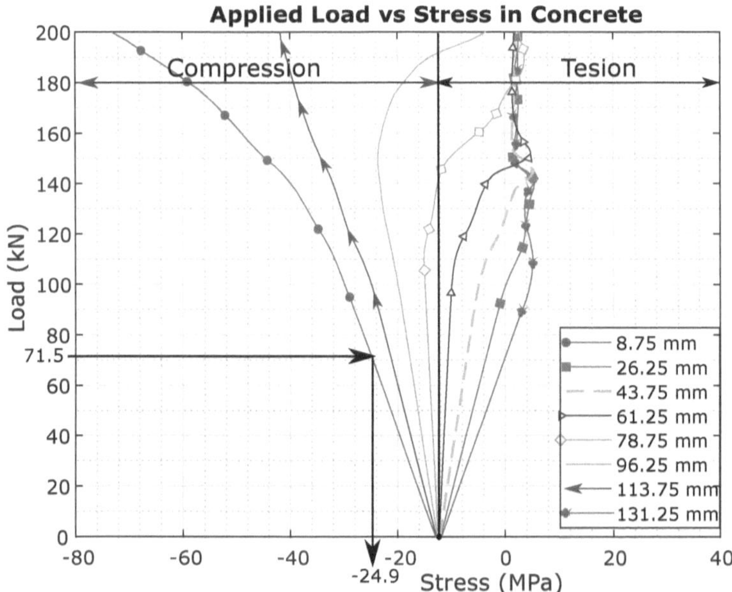

Figure 5.29: Maximum compressive stress in concrete as a result of 71.5kN applied load

As seen in Figure 5.30 number of cycles required to produce a fatigue failure in the concrete is 3.26×0^{11} this is an incredibly large number and the concrete can be said to practically never fail at this stress level.

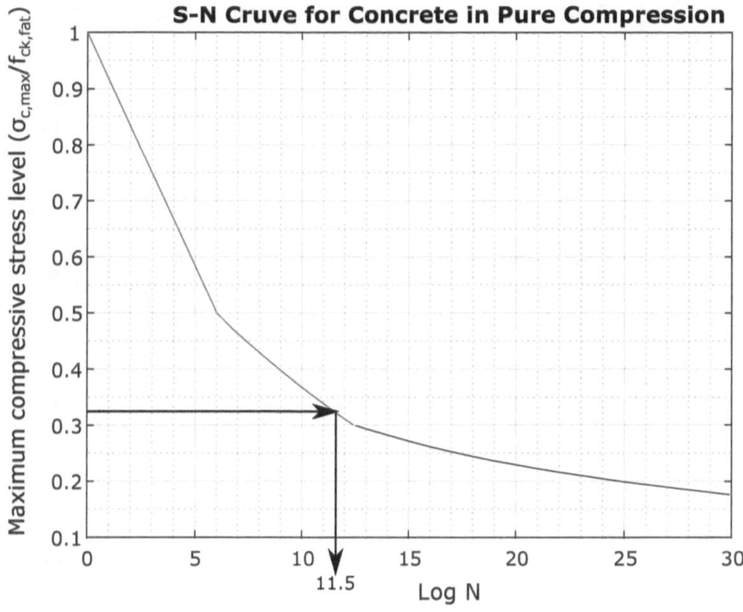

Figure 5.30: Maximum compressive stress in concrete as a result of 71.5kN applied load

Figure 5.31 shows the compressive stress required to cause failure at one million cycles and ten million cycles.

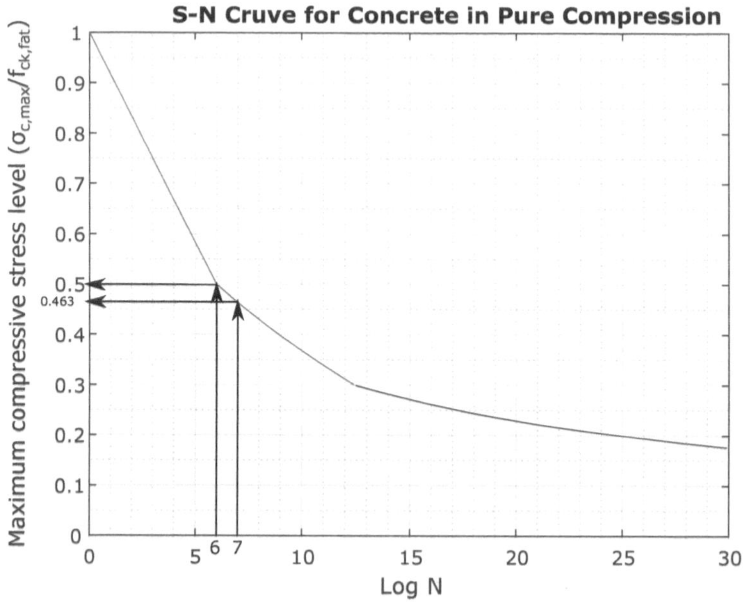

Figure 5.31: S-N curve for displaying compressive stress level relating to a fatigue life of 1 million and 10 million cycles

Table 16 contains the maximum compressive stress required to produce a fatigue failure at the respective number of cycles.

Table 16: Compressive stress level required to cause fatigue failure at 1 million load cycles and 10 million cycles

Number of Cycles	Compressive Stress level	Maximum compressive stress (MPa)
1 million	0.5	38.4
10 million	0.463	35.5

Using the Abaqus load vs concrete stress relationship as show in Figure 5.32, the
load required to produce a concrete stress which will cause fatigue failure at 1 million
and 10 million cycles can be found.

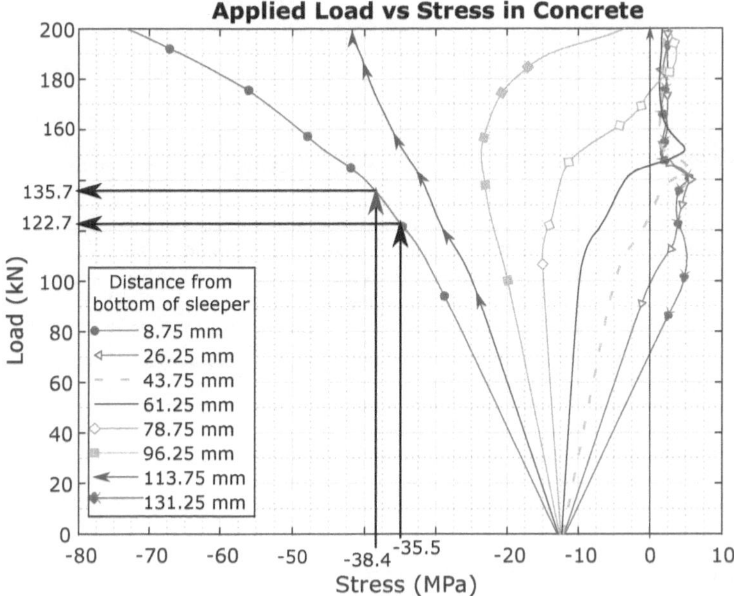

Figure 5.32: Load required to induce a compressive stress in the concrete that would
cause failure at 1 an 10 million cycles

Table 17 summarizes the required load to produce fatigue failure in the concrete for
one million cycles and ten million cycles these values are read off Figure5.32.

A summary of the fatigue calculations are shown in Table 19. Contrary to literature
which states that, the steel tendons fail in fatigue before the concrete it is evident
that the concrete is will fail before the steel. It must be noted that these results are
highly dependant on the S-N curves used to calculate them. The S-N curve for the
steel was modified to make it less conservative while the concrete S-N curve was not
modified from that specified in the fib code. To ensure that S-N curve predictions
match reality, physical fatigue test should be conducted on the sleepers.

Table 17: Load required to cause fatigue failure at one million and ten million load cycles

Number of cycles	Load required to produce failure (kN)
1 million	135.7
10 million	122.7

Table 18: Summary of loads required to produce fatigue failure at 1 million and 10 million load cycles

Number of cycles	Load required to cause failure in steel (kN)	Load required to cause failure in concrete (kN)	% difference
1 million	164.4	135.7	19
10 million	157.4	122.7	25

5.5.3 Validation of Austrian Standard Fatigue Test

According to Part F of (AS 1084.14-2012) the fatigue test should be carried out as follows:

1. Load the sleeper at a maximum rate of 25kN/min until structural cracks occur.

2. Release the load

3. Apply a repeated load for 3 million cycles varying from 15kN to $1.15P_2$

4. The loading frequency must not exceed 600 cycles per min or 10Hz

5. After the 3 million cycles have been applied the sleeper must be able to sustain a load of $1.15P_2$ for 3 min.

P_2 is defined as the test load required to produced the required rail seat positive moment in kN and is calculated according to the equation below.

$$P_2 = \frac{2M_{R+}^{cr}}{0.33 - 0.045} \tag{6}$$

M^{cr} represents the moment required to start tensile cracking

For the low profile sleeper P_2 was calculated as 81.6 kN. Thus $1.15P_2 = 93.87kN$. The detailed calculations are included in Appendix D.

Figure 5.33 displays the fatigue loading schedule that was specified by the Australian code and subsequently applied to the sleeper.

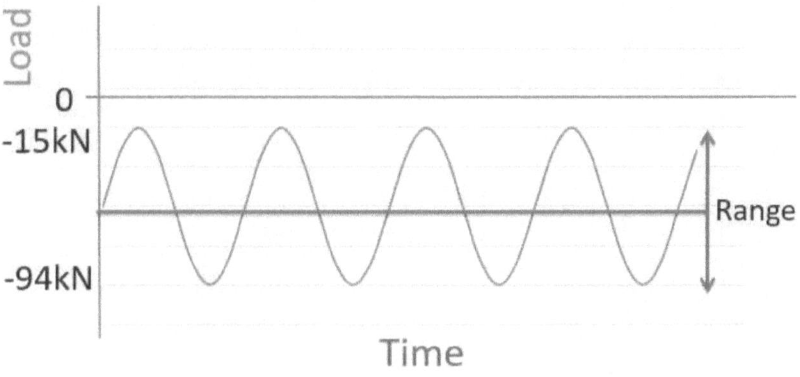

Figure 5.33: Fatigue loading schedule

5.5.4   Fatigue tests conducted in laboratory

The defined loading schedule was applied to two sleepers. As similar results were obtained for both sleepers only one set of results will be displayed. The fatigue test was conducted as follows:

A non-destructive static test was conducted by loaded the sleeper to 80 kN which is below the first crack value, the displacement and strains were recorded. The load rate was limited to below 20kN per min This non-destructive static test was conducted before the fatigue loading started and then every one million cycles thereafter.

The deflection of the sleeper and the strain measured on the top and bottom were measured for each non-destructive static test. These measurement provided an indication of fatigue damage progression. The Australian standard states that three million cycles must be applied to the sleeper. The loading schedule was extended to four million cycles to ensure that the standard had been adequately met an exceeded.

Figure 5.34 displays the displacement results. By observing the slope of the curves it is evident that the beam's stiffness did not deteriorate as the number of load cycles increased.

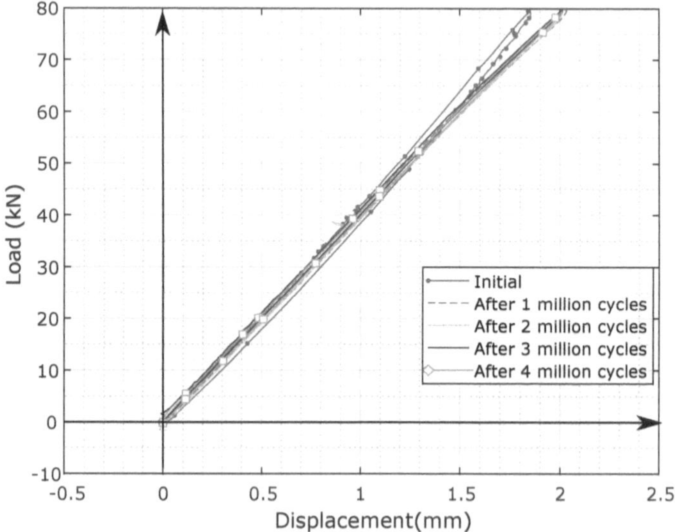

Figure 5.34: Midpoint displacement vs load measured after different numbers of load cycles

Figure 5.35 and Figure 5.36 display the strain measured at the bottom and top of sleeper respectively. As can be observed there is no increase in strain as the number of loads increases. The tensile strains curves are continuous which is indicative that the fatigue did not cause any cracking on the sleeper soffit.

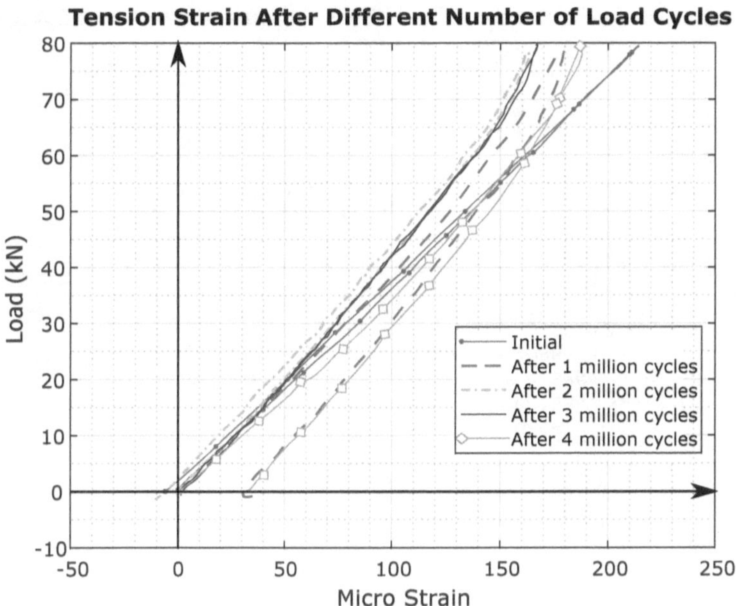

Figure 5.35: Tensile Strain vs load measured after different numbers of load cycles

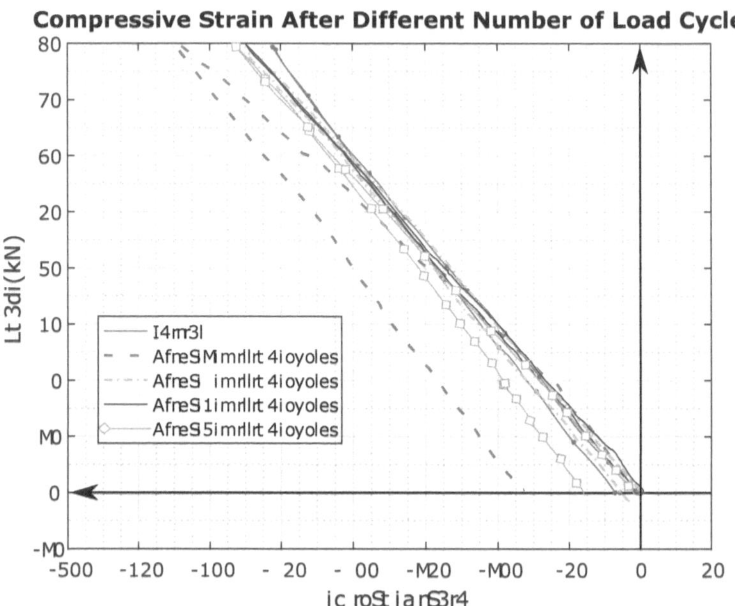

Figure 5.36: Compressive Stain vs load measured after different numbers of load cycles

5.5.5 Numerical Model Predictions

The FEA model is used to predict the stress in the steel when the sleeper is subject to a load fluctuating from 15 kN to 93.87 kN.

As can be seen in Figure 5.37 the maximum stress is found in tendon 4. The stress in this tendon associated with the loading of 15 kN and 93.8kN is 1258.4 MPa and 1290.5 MPa respectively resulting in a stress range of 32 MPa.

123

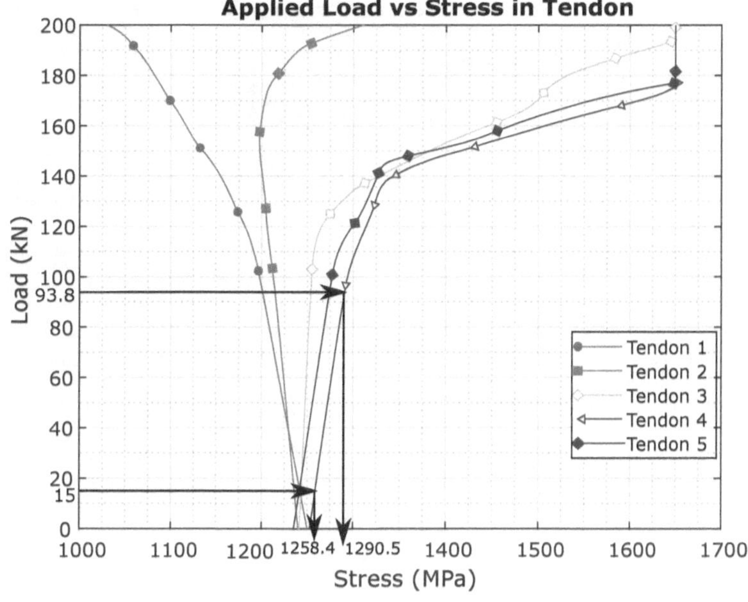

Figure 5.37: Tensile stress range resulting from Australian fatigue test

With a stress range of only 32 MPa it is not worth trying to plot the fatigue life of an S-N curve as this stress range is well below the fatigue endurance limit.

The compressive stress relating to the applied load of 93.8 kN is found using Figure 5.38.

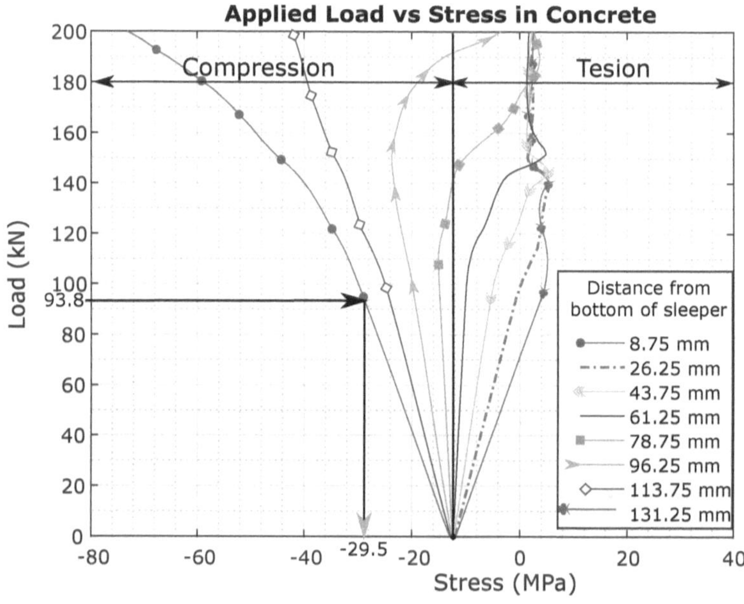

Figure 5.38: Maximum compressive stress resulting from Australian fatigue test

When this compressive stress is converted to the maximum stress level and plotted
on Figure 5.39 the number of load cycle until failure can be identified. At this stress
level the theoretical maximum number of loads before failure is one thousand million
cycles.

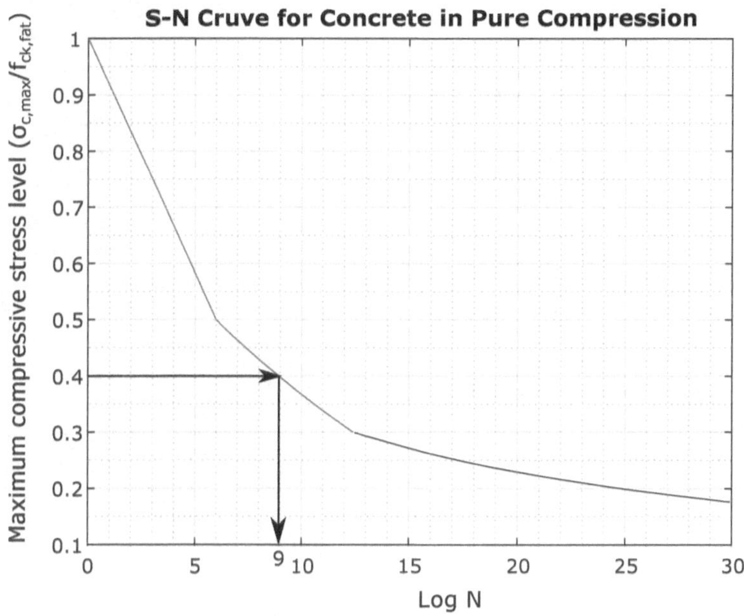

Figure 5.39: Concrete S-N curve displaying number of loads relating to the Australian fatigue test

This numeric study explains why there was no visible deterioration in the laboratory tests. These findings bring the usefulness of the Australian standard fatigue test into question. It is evident that under the current loading schedule the stress range in both the steel and concrete is far to low to expect any fatigue deterioration ever let alone within the specified three million cycles.

6 Conclusions

A global shortage of hard wood sleepers and a desire for better track stability is driving the replacement of wooden sleepers for pre-cast concrete sleepers. Low Profile concrete sleepers have the same vertical dimensions as wooden sleepers and are thus an ideal replacement choice however a better understanding of the load response especially the fatigue behaviour is necessary.

Like most pre cast concrete members, Low Profile sleepers are designed according to the minimum stress criteria where the maximum tensile stress is limited to 3.5 MPa. Low Profile sleepers are required to satisfy this criteria while sustaining a 11.8 kNm bending moment. Modern design philosophies such as the fatigue limit state offer a more economic use of materials and are thus widely used in structures such as bridges. The potential to use a fatigue limit state in the design of Low Profile concrete sleepers was investigated in this study.

The study employed both physical laboratory tests and numerical modelling. Three Low Profile sleepers were tested to failure in a 4 point flexural test while two sleepers were subject to a 4 million cycle fatigue test. The effect of impact loads was not investigated. A standardized test procedure specified by the Australian (AS 1084.14-2012) code was utilized. According to this standard the concrete sleeper was tested in flexure, isolated from other track components.

The commercial finite element program Abaqus was used to develop and process the numerical model of the low profile sleeper. A concrete damage plasticity model (CDP) was chosen to represent concrete's stress vs strain behaviour while a purely elastic-plastic model was selected to represent the steel tendon stress vs stain relationship. The FEA was tuned by comparing the mid point displacement between the FEA model and the monotonic physical lab tests. The five variables that were calibrated in order to tune the FEA model were; •

- Elastic modulus of concrete

- Dilation angle of concrete

- Kc (parameter in yield criterion)

- Concrete compressive strength

- Pre stressing force

Fatigue was not modelled directly using the numerical model. The model was used to obtain the stress field for the both the steel and concrete. The stress field was related to the fatigue life via S-N curves provided by fib Model Code 2010. An amended steel reference strength of 300 MPa was used instead of the 185 MPa value as stated in the fib Model code 2010. This was done in an effort to achieve a less conservative and more realistic S-N curve.

The average static failure of Low Profile sleepers occurred at 198 kN. At this load significant cracking occurred and the average midpoint displacement was 27.65 mm. Low Profile sleepers can be classified as under reinforced pre stressed beams and typical failure for such members was observed. Flexure failure resulted from significant steel yielding before concrete crushing occurred.

The load at which concrete first displays macro cracking is an important value as it is a load referenced in the the Australian standard and concrete member durability is drastically reduced due to the presence of macro cracks. The first crack load was predicted by both the FEA model and standard elastic beam equations as 98 kN and 88 kN respectively. This is compared to an average first crack load of 85.3 kN measured in the lab. It must be noted that both the FEA and elastic equation predictions are highly dependant on the assumed tensile strength of concrete. A better match between the measured values and predicted values could be achieved if the assumed tensile strength was reduced, however in an effort to follow the Australian code the assumed tensile strength was obtained from the prescribed formulas.

Using the FEA stress field in conjunction with the fib Model 2010 S-N curves the following fatigue results were obtained;

When the sleeper was subject to a fatigue load fluctuating between 0 kN and 71 kN (the load associated with the maximum allowable stress design bending moment of 11.8 kNm), the steel stress was under the fatigue limit while the concrete had an predicted fatigue life of $3.26x10^{11}$. Practically speaking it can be stated that at this load regime the sleeper will not fail in fatigue. The maximum stress design criteria is thus excessively conservative when compared to the calculated fatigue capacity of the low profile sleepers. It must be noted however that the maximum stress design ensures that no cracking occurs and the durability of the sleeper is maintained.

Table 19 contains a summary of the 1 million and 10 million cycle life study.

Table 19: Summary of loads required to produce fatigue failure at 1 million and 10 million load cycles

Number of cycles	Load required to cause failure in steel (kN)	Load required to cause failure in concrete (kN)	% difference
1 million	164.4	135.7	19
10 million	157.4	122.7	25

Contrary to literature, which states that the steel tendons fail in fatigue before concrete failure, it can be seen that concrete shows failure at loads lower than the steel tendons. This deviation from literature can most likely be attributed to different S-N curves being used. It is valuable to restate that for this study the S-N curve for steel was modified to make it less conservative. If the original S-N curve was used the steel would be predicted to fail before the concrete in accordance with literature. The importance of using an S-N curve that matches reality rather than a conservative design curve is highlighted by this observation. The difference between steel and concrete failure loads is less for failure associated with 1 million loads than 10 million loads. This indicates that at high stress ranges the steel and concrete have a similar fatigue life.

There is only a 7 kN or 4% between the load required to produce a steel failure at 1 million cycles and the load required to produce a failure at 10 million cycles. This indicates that once a critical load limit has been reached in the sleeper, fatigue failure becomes increasingly sensitive to small load increase. This sensitivity makes the fatigue failure difficult to predict at high loads.

The fatigue test as prescribed by the Australian code (AS 1084.14-2012) requires Low Profile sleepers to be subjected to 3 million load cycles ranging from 15 kN to 95 kN. At this load regime the steel tendons experienced a maximum stress range of 32 MPa which is below the steel fatigue limit thus steel fatigue can not be expected. The maximum concrete compressive stress experienced for this load level was 30 MPa which theoretically results in a one thousand million fatigue life. These results bring the usefulness of the Australian fatigue test into question. Under the specified load regime neither the steel nor the concrete are predicted to fail in fatigue let alone show signs of damage within 3 million cycles.

While this study indicates that the fatigue capacity of Low Profile sleepers allows for significantly higher loads than allowed by the maximum stress design criteria, at loads higher than 85 kN cracking occurs. The sleepers life is thus governed by a serviceability design state rather than a fatigue limit state.

www.ingramcontent.com/pod-product-compliance
Lightning Source LLC
LaVergne TN
LVHW041702190726
843493LV00007B/1913